# ADVANCED INFORMATION TECHNOLOGY IN EDUCATION AND TRAINING

# ADVANCED INFORMATION TECHNOLOGY IN EDUCATION AND TRAINING

Arthur Cotterell and Richard Ennals,
with Jonathan Harland Briggs

Edward Arnold
A division of Hodder & Stoughton
LONDON BALTIMORE MELBOURNE AUCKLAND

First published in Great Britain 1988

*British Library Cataloguing Publication Data*
Cotterell, Arthur
Advanced information technology in
education and training.
1. Education — Data processing
I. Title II. Ennals, J.R. III. Briggs,
Jonathan Harland
370'.28'5 LB1028.43

ISBN 0 340 40678 X



Typeset in Linotron Plantin by Taurus Graphics
Printed and bound in Great Britain for Edward Arnold, the educational, academic and medical publishing division of Hodder and Stoughton Limited, 41 Bedford Square, London WC1B 3DQ
by Richard Clay Ltd, Bungay, Suffolk

# Contents

# Preface

The argument of this book is that advanced information technology is currently particularly amenable to use in education and training, not just for their benefit, but as an essential stage in the development of the technology itself.

Computers and computing have a relatively short history of practical use, during which they have become the preserve of specialists who have devoted considerable time and energy to mastering the intricacies of the technology. Busy teachers, lecturers and trainers in non-computing subjects have waited for the technology to deliver its promised benefits in a manner that might benefit education and training in their area, and have largely been disappointed. Some have been prepared to compromise their specialism in order to exploit the technology, and have formed part of the cottage industry of educational software production. After months or years of effort it has become apparent that the results do not match up to initial hopes and expectations. In many cases disillusion has set in, and computer hardware and software now gather dust in cupboards.

The general public whose expectations of a computerised society were built up over the last two decades may consider that computer science has failed to deliver all it promised. Human problems do not appear to be such that computer systems can reliably solve them.

In education and training we do not aspire to perfection: we seek to enhance our understanding and that of our students and trainees, equipping them to deal better with problems in the real world. Education and training are concerned with preparation, with the provision of sound structures to support thought and action, both through the acquisition of particular specialised skills and through the development of successive levels of ability in abstraction, generalisation and problem-solving.

There is a real sense in which computers and computing have come closer to the position of education and training. It has become clear that about forty years of computer science and computer systems have not solved the world's problems, but that we may have a more acute perception of those problems, and of our degree of inability to solve them at present. Whereas it was fashionable in the 1960s to say that computers had come of age, we are now perhaps aware of the distinction between adolescence and maturity. At the age of fifteen, students often feel they know all the answers; by the age of forty they may realise that they only know some of the questions.

Advanced information technology, and in particular fifth generation computer systems now under development, do not represent a simple continuation of the trends in computing to date, but a marked change in direction in a number of respects. The realisation by computer professionals that their traditional training may now be irrelevant is a cause of instability. Who could have imagined that computer programmers would face redundancy or retraining? Their fate symbolises the increasing speed of technological advance. In this respect it is not unlike the world of education and training, where we have always known that change is inevitable.

There are particular respects in which education and training can come to the aid of advanced information technology, and vice versa. 'Intelligent knowledge-based systems' need to be based on knowledge. 'Expert systems' presuppose the existence of expertise, available in an accessible and comprehensible form. 'Intelligent tutoring systems' require the designer to have an appreciation of intelligent tutoring, as well as considerable experience of the problems and behaviour of diverse tutees.

Complex computer systems now under development require understanding not just of computer systems but of their human users, working collaboratively to solve problems. Software engineering presupposes some understanding of software lifecycles: software, like students, needs to be provided with sound structures and the capability of adjusting to changing circumstances. In order to build a working system you have first to arrive at an understanding of the problem to be addressed. Recent research in computer science areas such as knowledge elicitation, dialogue systems and natural language understanding has shown how much has still to be learnt. Educationalists in the Western academic tradition have spent centuries addressing questions which are now facing computer scientists. Computer science can therefore offer some powerful new tools with which we can cast new light on old problems.

Fifth generation computers, once they are realised, will be used by our present generation of schoolchildren. Issues from within education and training have had a fundamental influence on the development of such computer systems. Education and training will clearly be affected by these systems in turn,

This book seeks to provide an intelligent and intelligible context for the new generation of technology in terms of experience of past generations, both of systems and of education. Focusing on practical applications in education and training today, and expressed in non-technical language, it seeks to demystify what should be a liberating, enabling technology, available to all. It is based on extensive practical experience in the United Kingdom and internationally, and in particular at Kingston College of Further Education.

# 1 The Evolution of the Computer and Computing: First to Fourth Generations

## 1.1 Automating reasoning

Although working computers have only been in existence for some forty years, the history of attempts at the automation of human reasoning stretches back for millennia. One example is a very ancient instrument devised to aid mental arithmetic, the abacus, whose origins probably lie in West Asia. Today the pocket calculator is in fact only beginning to replace it in the commerce of many countries.

The first computers were indeed a physical realisation of an ambition that had previously been expressed in the construction of the abacus, which in its most developed form employs sliding beads on a frame. Numbers have long had a special fascination, offering as they do the means of abstract manipulation divorced from a context of meaning. We are accustomed to learning arithmetic, not just for its own sake, but because the numerical operations at which we become proficient play an important part in practical aspects of life. Many subjects are not naturally expressed in numerical terms, but we have come to realise that there can be real advantages in using figures and quantitative methods, rather than simply relying on general qualitative description.

Other technologies have evolved to assist in handling and processing different forms of information; for example, printing has made knowledge expressed in words available to a mass audience. Language permits communication between individuals, who agree to common meanings for the words and symbols that are used in discourse, though misunderstanding is such a common occurrence that philosophers have been obliged to devote much of their energies to the definition of terms. Writing enables something to be expressed once, and is available for later consultation or reference. This has been mechanised with devices such as typewriters, and more recently word processors. Human capabilities are extended by such representations of organised human thought. Books and letters are examples of forms of knowledge representation which have become accepted as part of civilised life, and with printing, mass communication is possible. Communication can take place with people that one has never met, even across the generations from those already dead, and to those yet unborn. With the technologies of broadcasting the written and printed word can be bypassed, allowing less formal modes of expression, and greater immediacy. There is increasing experience of mixed media communication, using for example a combination of text and audio, or text and graphics, sometimes united through television, recorded video or interactive video devices. Communication thus becomes dynamic and interactive, and can be more conversational than didactic.

Despite developing from a wealth of technologies, the computers built so far, the first four generations of computer systems, can only provide a partial realisation of the dream of automating human reasoning. There are a number of respects in which we can now see this is the case, implying no criticism of the innovative work in the field, but rather expressing admiration at what has been achieved with such impoverished technological and intellectual tools.

Using these computer systems has involved an inevitable compromise because people have had to learn to think in the manner of a machine so as to gain its assistance in solving their problems. The first computer users were compelled to use machine language, and give instructions, in the binary code of zeros and ones directly used by the electrical circuitry within the machine. Successive research advances have produced higher-level languages that more closely approximate to natural human language but without attaining its power, flexibility and natural correspondence to the way humans think. Just as philosophers argue that the limits of language coincide with the limits of the knowable world, so the world of individual experience seems unduly limited if it is circumscribed by the linguistic constructions of a computer language such as FORTRAN (available from 1957) or BASIC (which followed in the 1960s).

This problem goes deeper even than language. For practical reasons computers have so far followed the same principles of underlying design, or *architecture*, as laid down by the mathematician John von Neumann (1903–1957), who built in the United States some of the first working systems immediately after the Second World War. Fundamental to this design is the principle that computers have one central processing unit, which can deal with a single stream, or sequence, of instructions. This has been reflected in the computer languages that have been developed to use such systems, so that the order of instructions is of critical importance. The sequence in which operations are to be performed is enormously significant. This is even exhibited in approaches to algorithms (methods for problem-solving to be followed by people or machines). Just as a person can only type one sentence at a time, there is the working assumption that computers can only do one thing at a time.

Human beings, of course, are not like that, and it is a distortion of our way of thinking, and of our approach to problem-description and problem-solving, to assume that we simply operate sequentially in the manner of a computer.

Another consequence of increased involvement with conventional computer systems has been a requirement for certainty, for accuracy to the point of being spurious. The myth has arisen that human problems have definite answers, rather like simple arithmetic; it has been imagined that descriptions can be absolute and that problems in the real world can be solved by man-made technological systems.

As a result, conventional computing has tended to be culturally divisive. Computer programmers have been able to live a hermit-like existence, communicating only with machines in a world devoid of human contacts. Elegance and conceptual economy could thus prevail over more normal human criteria, such as cultural appropriateness, cost and comprehensibility. In education the concentration on numerical methods and a mechanical sequential approach has proved unattractive to subject specialists who were not prepared to make the compromises that they felt were required of them, at the expense of their subject.

For technical reasons computer systems have been difficult for humans to use. Typically input has been via a typewriter keyboard in an unnatural, formal manner, and output has been on reams of paper. In recent years advances in the way in which humans can interact with technology ('human-computer interaction' or the 'man-machine interface') and the progressive integration of related technologies are making easier communication possible.

These developments promise to make our use of computers more straightforward. However, to increase their problem-solving powers, their reasoning strategies must be improved; they must be made to mimic human problem-solvers but human reasoning is complex and poorly understood. The successive development of computer systems, from first generation to the current fourth, has made it clear that we cannot give a full and adequate description of our own reasoning. Since we are unable to describe our reasoning processes to ourselves or to our colleagues, we should not be surprised that we cannot describe them to a computer well enough to enable a replication of our reasoning whenever the machine is required to go beyond the information given.

In order to perceive the limitations of the first four generations of computers we also must have an appreciation of what has been achieved. Many practical and large-scale problems have been solved where they conform to the structures and means of expression offered by the computer. We have also come to a greater understanding of many problems through failing to solve them: if nothing else, the concept of debugging a program has been usefully carried over to debugging one's own ideas. We have been required to give full and unambiguous descriptions of our problems in terms that the machine can understand. We should not be surprised that we have often failed, nor should we be disheartened.

## 1.2 Patterns, processes and rules

The idea of pattern is fundamental to human understanding of a complex world and of complex processes. Patterns enable us to simplify both the description and comprehension of a problem, and to suggest to others the processes by which they can be solved. Once we are able to identify a common pattern in separate problem areas we can then focus on essential differences of detail, and make informed comparisons. All of us have experience of using maps and diagrams in this way, even if it is merely comparing the relative virtues of new models of cars displayed in the sales literature.

In the eighteenth century, during the first industrial revolution, attempts were made to mechanise industrial processes, thereby extending human muscle power with the power of machines. The textile industry was an early customer for steam engines, used to drive new machines for spinning and weaving. Attention was paid to patterns, with the objective of increasing production and efficiency. In France, Joseph-Marie Jacquard (1752-1834) developed an automated loom that was controlled by perforated cards; these early 'punched cards' gave instructions which orchestrated the machine's entire operation. Though primitive by modern standards, the idea was significant because weaving machines, with their diverse mechanical components, work in parallel and in several dimensions to produce complex patterns in textiles following descriptions by designers and management. Through a common pattern, made into a common program, the different elements of the process could be made to cooperate, while making best use of scarce design expertise.

This idea of punched cards was taken up by Herman Hollerith (1860-1929), who constructed a machine to help in the compilation of the 1890 American census. It was quite apparent that a mass of information was embodied in these census returns, to which access could be provided easily through the discernment and encoding of a pattern. The potential of this knowledge representation was restricted by the technology itself since the sorting of cards was performed sequentially. It took a considerable time to search for a single pattern and in the process produced far more information than was wanted. Because early computer technology could only cope with a crude form of searching — checking for the presence or absence of a characteristic or with numerical coding — so much of the richness of the knowledge embodied in census returns was lost. Using a present-day *relational database* or *logic database* (where the computer representation of the knowledge is organised in a way that more closely resembles the way we are accustomed to organising knowledge in tables), the same knowledge captured in the census return can be made to yield far more detailed results, in response to a greater range of questions. A number of educational computing packages are now available which exploit this advance in technology, making old demographic wine available in new bottles. Using a package such as QUEST, a school student can investigate the history of his or her own area, drawing on census returns and trades directories.

Patterns and processes are of fundamental importance in manufacturing industry, as visionary entrepreneurs have long known. Josiah Wedgwood (1730-1795), founder of the pottery firm which still bears his name, revolutionised both the production and marketing of china. He emphasised the important of pattern, high standards and systematic processes at each stage of his business. His essential skill lay in identifying the customers' requirements, finding or commissioning a design that would appeal, and building successive prototypes using different materials until he produced a finished item that would command attention.

There is also a historical tradition of studying management and decision-making in terms of patterns and processes, long before the sociologist Talcott Parsons and his colleagues at the University of Chicago identified them in the working of society. Niccolo Machiavelli (1469-1527) sought to find a successful pattern of conduct for the rulers of Italian city-states. He inferred rules of politics and warfare, which as an adviser he could pass on in the form of advice to his princely clients. His advice on the running of complex city states has much in common with the advice of modern consultants on the running of a project, whose skill is to discern patterns and regularities amid complexity and thereby extract rules.

Patterns can assume many forms in different media and contexts. Languages can be taught with an emphasis on patterns, as games with rules which can be learnt; someone has learnt the rules of a game when they know how to make the next move. Literary critics delight in identifying stylistic characteristics of different writers, enabling them to write pastiche, or parodies. Architecture and town planning can be seen in a similar light; a designer may choose a classical or georgian approach. The training of artists and musicians typically includes the development of a facility in producing work of a particular pattern.

At the lowest level patterns can be binary codes, describing a pattern of electrical connections in a machine. One way of viewing the last forty years of computing is in terms of bringing such patterns nearer to ordinary people. Assembly languages, for example, allow the programmer to use more familiar words and phrases instead of instructing the computer with binary codes or hexadecimals.

Patterns used to represent knowledge are vitally important to work within the field of *artificial intelligence*. Instead of manipulating numbers the computer systems of the future will manipulate symbols and patterns. The new software known as *expert systems* often uses textual rules to store knowledge: these rules have a fixed pattern which permits them to be manipulated easily. For example, here is a rule from a simple medical expert system:

```
Patient should take asprin if patient has
headache and patient does not have
sensitive stomach
```

The rule consists of phrases linked by 'if' and 'and'. This pattern is important. When the computer is asked a question such as:

```
Fred should take something?
```

The computer matches the question to the rule before providing the answer. It might be:

```
If Fred has headache and Fred does not
have sensitive stomach. Then Fred should
take asprin.
```

This process is called *pattern matching.*

## 1.3 Breaking codes

War, or its imminent threat, has often stimulated technological advance. Therefore, it is not surprising that computing was also a product of conflict this century.

In both Britain and the United States the Second World War drew together talented individuals on military projects which ultimately had wide-ranging benefits for mankind. A remarkable group of mathematicians was gathered at Bletchley Park in Gloucestershire, with the mission of breaking the codes used by the Germans and their allies through their Enigma machine, knowledge of which was acquired during the invasion of Poland. Cryptography was thus forced ahead, computer systems being among the unexpected by-products. Among the original Bletchley Park group were Alan Turing, Donald Michie and Jack Good, all of whom have played leading roles in the subsequent development of artificial intelligence.

On the other hand, John von Neumann was always more directly involved with the military, serving even before the declaration of war by the United States, as an adviser to the Ballistics Research Laboratory in Maryland. One of the problems put to the laboratory was the calculation of shell trajectories; it led to the first electronic calculating machine being commissioned in 1946.

Cryptography represents, in a formal sense, a paradigm for complex computational problems. Often the problem can be expressed in terms of translating from 'machine code' to 'human code' or natural language, or vice versa. Translating may be between artifical computer languages or between natural languages.

Techniques developed in computer science have cast new light on the nature and structure of natural language with the production of interpreters, which handle individual sentences as they are entered by the programmer, and compilers, which deal with whole programs, translating them into a language understood by the machine. Noam Chomsky, working in the field of linguistics, developed approaches known as structural and transformational linguistics which have had a profound effect on subsequent work on both artificial and natural languages: they have in effect created the new field of computational linguistics. On being asked to explain the origins of PROLOG (PROgramming in LOGic), which was first available in Marseilles in 1972, its implementer, Alain Colmerauer, pointed to the work of Chomsky. This new computer language was adopted by the Japanese in 1981 as the starting point for their Fifth Generation Computer Project, which has set the world rethinking about computer design and function.

Colmerauer had worked in Montreal on automatic translation between French and English, where Prime Minister Trudeau's political speeches were needed simultaneously in both languages. Despite early optimism, automatic translation turned out to require a deep level of understanding of both the syntax (the grammar and form) and the semantics (meaning) of language, rather than just brute force use of dictionaries. Colmerauer also worked in Grenoble on compilation techniques for computer languages, and found that the formal notations of logic played a crucial role in the processing of both natural and computer languages. He collaborated with Robert Kowalski, whose research has been concerned with the use of logic as a computer language, developing the approach known as *logic programming.*

A similar approach to problems of understanding human language and action can be seen in linguistic philosophy. Understanding a picture, or a complex set of behaviours, can be, according to Ludwig Wittgenstein, like 'making a move in a calculus'. His lectures at Cambridge University, many of them published after his death by his former pupils, have had a wide influence. Philosophers distinguish between what people mean *in* saying something and what they mean *by* saying something. These are known as the illocutionary and perlocutionary forces, and this style of analysis is now applied to actions as well as to 'speech acts'.

In short, progress in computer science research in the identification of different levels of language, abstraction and translation that are involved in mastering codes can have far wider signficance. Even if problems are not solved we can arrive at a clearer understanding of the levels at which difficulties arise. For the practical use of such ideas and techniques in real problems of knowledge, it must be possible to move beyond numerical and alphanumerical codes into the use of language with more meaning in relation to the real world. We would like to be able to use the richness of natural language, and the power of the system to help us to grapple with shifting representations of knowledge. As human beings we have learnt to cope with enormous problems of language; we must pass some of our understanding on to machines if we are to be free to tackle new problems and crack new codes.

## 1.4 People thinking like machines

The first four decades of computer use have involved inevitable compromises. Users have been at the mercy of systems programmers, who have developed computer languages to a large extent dictated by the processors supplied by hardware manufacturers.

The new trend in computer design, known popularly as *fifth generation computer systems*, is moving towards the development of *declarative systems*. This offers the prospect of an increasing role for the thinking user, who is made less dependent on the past decisions of microprocessor designers. The responsibility of the intelligent user is to describe the problem in a chosen way that is taken as input to the computer system. This description of the problem is a major factor in determining what computer language to use, and what model or approach to computation is required. This in turn determines what combination and configuration of processors are employed in solving the problem.

We can never expect complete user-control in the solution of complex problems through computer systems; there will always have to be some degree of compromise by the user to take account of the limitations of his mechanical partner. However it has become easier to address a problem with the aid of a computer, without being wholly dominated by the mode of operation of the computer.

'Thinking like a machine' is interesting in the sense that it implies a process of reflection upon the process of thinking itself. In seeking to think like a machine with a single processor and a very limited vocabulary we are obliged to enter a new artificial world, where what we can say is tightly constrained by the tools with which we can express ourselves. Our expression is always limited by our language, but this artificial constraint can oblige us to think of old problems in new ways. Early computers had very little memory, and were extremely expensive to run, thus placing pressure on programmers to develop programs that required the minimum of memory and processor time. More recent personal workstations offer generous memory facilities and support artificial intelligence computer languages that are very demanding of space, because they have been designed with an emphasis on the form of problems to be solved rather than with the objective of complying with the details of current computer design. They may be less efficient in terms of speed, but are more likely to be comprehensible to the programmer and a third party user, as they can be expressed in almost natural language, and can be understood wherever they appear in a complex program.

The advance of technology and research has given us new insights into the tasks that we can expect to be performed by machines. The traditional view was that machines would take over those tasks that are regarded as menial or mundane when performed by humans. In the industrial revolution, workers in textile mills became merely 'machine minders'. This idea carried over to many other stages in industrial processes, some of which, such as spot welding in automobile plants, can now be undertaken by robots. Whole factories and assembly plants can be seen running without human hands, as long as the process is running precisely according to plan.

Problems arise when machines are required to interact with the real world. It is far from trivial to teach a machine to sweep a floor, particularly if we expect it to clean around obstacles or to distinguish between types of object. We may pay a cleaner a low wage to clean ashtrays in our offices, but it is beyond the state of the art in the forseeable future to expect a robot to distinguish reliably between an ash tray and an antique Ming vase. In museums for example, mechanisation could prove to be an expensive mistake.

To the alarm of some professionals, who had felt secure in their exclusive status, it turns out that the work of many professions is more amenable to automation than is much manual work. If you ask for advice from a bureaucrat behind a desk, he may provide it in a 'mechanical manner', reaching for the appropriate rule book or set of regulations, and reading out the relevant sentences, often charging you a large fee in the process. To the extent that it is always the same rule book that is appropriate, the bureaucrat could be automated. To the extent that his individual judgement or interpretation is required on separate cases, his human job may be safe.

In many cases there is virtue in acting with machine-like precision; we like brain surgeons to be precise and consistent, though we may be reluctant to hand over their tasks to an unaided machine. We may pride ourselves on our skills in shaving in the morning, and feel reluctant to submit ourselves to shaving by a robot, which cannot be relied on to avoid causing us pain.

There are often external environmental factors extrinsic to the description of the task itself that govern our decision as to how it should be undertaken. We might, for example, be happy to entrust the well-understood task of sheep shearing to an automatic device, given the current evidence that the task can be performed without cutting and hurting the sheep. We might remain reluctant to replace hairdressers by robots. This special case points up the general position; it is a matter for human management decision, possibly at the social level, as to which tasks should be performed by people and which by machines. We might reach some criterion such as whether performing the task is pleasant or unpleasant for a person, or whether it exceeds a certain degree of risk or expense. The answers will vary between situations and over time, according to prevailing circumstances of manpower, economics, urgency and ethics. Scarce human skills which are widely needed may require the assistance of machines, simple tasks such as preparing a meal may be easy to automate but may have some valuable social functions such as providing an activity in which a family may cooperate. Fast food is not always best.

The presence of all-purpose machines which can be programmed to perform a variety of tasks can be seen therefore as giving new life to old issues regarding work and economic behaviour. Computers provide a new complicating factor for economic, social and political theorists, whose language of discourse will have to change to reflect their presence. Much of the work of the theorists, to the extent that it is consistent and predictable, is itself susceptible to automation. Theorists who work as expensive consultants may see this as a threat to their livelihood. The Luddites of the 1990s may well wear pinstripe suits.

## 1.5 Thinking by numbers

Numbers have long commanded respect from humans concerned with thinking about diverse problems. They offer the apparent prospect of precision, though the choice of numbers to use in a given circumstance may be more arbitrary than many care to admit. Statistics has acquired great prestige for this reason; sponsors of research are often happier to receive results in quantitative form, with reports on standard statistical tests, and with measures of confidence, than to receive a statement in ordinary language, which appears less precise.

There is a choice. We can use crude tools to which we then assign numerical precision, or we can use the refined tools of language which defy quantitative expression. For example, if we are conducting an opinion poll, we can ask a simple question:

'Do you support the Prime Minister?'

and count the responses for and aginst. As a finer measure of opinion, we could ask:

'On a scale from 1 to 10, how would you rate your level of support for the Prime Minister?'

Approaching closer to the use of descriptive language, we could ask:

'Which of the following descriptions would you apply to the Prime Minister?'

beautiful
intelligent

good listener
flexible
compassionate
loyal

An alternative question might ask the respondent to give a numerical rating for each of these qualities, as before, or might ask the respondent to locate the Prime Minister on each of a series of continua, such as:

firm (1) — flexible (10)
dry (1) — wet (10)
strong (1) — weak (10)

An open-ended question, producing diverse responses which might defy easy quantitative analysis, would be:

'How would you describe the Prime Minister?'

Conventional computer science and statistics have chosen the first options, because their available technology has supported complex numerical operations. New systems may now encourage the latter choice, as we gain sophistication in machine-aided description, and lose confidence in numbers that cannot be explained in ordinary language. Mathematics is about more than numbers: it is concerned with the manipulation of symbols and abstract structures. We are beginning to understand how logic and language can be subjected to processing in a manner that we are more accustomed to use with numbers.

We could image using an enhanced word processor to strengthen our adjectives at a stroke — wherever we have written 'bad', it could be replaced by 'awful', wherever we have written 'boring' it could be replaced by 'incredibly tedious'. Such a device could be employed to speed the process of writing end of term reports. We could go a step further if our program knew the sex of the student, his or her favourite subjects and interests, and we could produce flowing sentences of appropriate content. A further extension would be to produce the report in a different natural language if a language other than English was spoken at home. Just as end of term reports are not usually in full English prose, they could just as easily be produced in fractured French, Italian or Urdu.

New generation computing is increasingly mathematical and logical in nature, as we learn to take advantage of the inherent properties of formal representations. Detailed study of a subject area, whether it is cricket, brain surgery, Zen Buddhism or motorcycle maintenance, often involves the learning of certain formal terms and notations, which could in turn be taught to the computer. We could imagine a computer system that was familiar with the rules, language and notation of tennis or football being of considerable use at the time of the Wimbledon Tennis Tournament or the World Cup. We are beginning to manipulate relationships between objects and equations as we have previously handled numbers, and in a consistent logical framework.

Much more work lies ahead, and we already know from the results of research in logic that there are inherent limitations to what is possible. Truly rich knowledge areas, such as cricket, lend themselves both to complex statistical analysis of run-rates, season and career first-class batting and bowling averages, size of wicket partnerships and other useful numerically expressed elements, and to elegant descriptive prose from writers such as John Arlott and E.W. Swanton. The recent England cricket captain Mike Brearley has even applied the techniques of philosophy and psychotherapy to cricket. We can envisage computer models of the Australian cricket team used to help in the training of future England batsmen and bowlers in their attempt to win the Ashes. But somebody still needs to know how to play, and this knowledge is not reducible to formulae of a numerical or any other nature.

## 1.6 Structure and abstraction

We live in a world of great complexity, and psychological research indicates that there are limits to the amount of knowledge that can be manipulated by the human mind at any one time. We have to learn to be selective, to identify common structures in different problems so that we can apply lessons that we have learnt elsewhere. We have to learn skills of generalisation and abstraction, so that the knowledge we are manipulating has as great a power as possible.

The computer is a uniquely valuable tool in this context, for it gives concrete forms to abstractions, enabling us to deal with concepts in the form of programs. When we have conquered the complexity of a small problem, we can capture our resulting understanding in a program which thereafter embodies that understanding, and makes it available whenever it is needed. We can then clear our minds for other problems, other areas of complexity, or higher levels of abstraction when we reason about programs themselves.

It has long been argued by computer science theorists that the skilled programmer needs to be able to cope with abstraction. This applies both when considering the nature of a problem of knowledge and when dealing with its mode of solution by a computer. In the early years of practical computer science the emphasis has been on increasing the power and sophistication of computer solutions, developing new and more effective tools and techniques. Computer languages, for example, were first written in a relatively *ad hoc* manner, determined almost wholly by the internal workings of the machine. Progressively, with languages such as ALGOL (ALGOrithmic Language), devised in 1958, and PASCAL, named after the French philosopher and devised in the 1960s, there has been more of a concern for clear formal definition, so that elements of a program can be regarded with increasingly mathematical precision. The power of computer variables has been appreciated, so that a single statement with variables can be made to apply in a variety of circumstances while remaining consistent and understood. Languages have provided tools with which to write structured programs, with standard computational structures derived from the advancing theory of algorithms. Programs have themselves begun to reflect in their structure the perceived structure of the problem to be solved, with the advance in 'systems thinking' and a concern for process as against content or product.

More recently it has become apparent that it is not sufficient to have an efficient program, a set of procedures for solving symbolic problems of a given surface appearance, transforming a set of input data into output data. We must also give attention to the nature of the problem of knowledge, to data description and to identifying the facts and descriptive generalisations or rules that enable us to come to an informed understanding of the problem.

Often a problem turns out to be soluble in a straightforward manner once it is clearly described. Without clear problem description there cannot be genuine solutions. Concepts and ways of thinking derived from advanced computing research can be applied to the stages of problem-solving which precede the writing of the code of a program which is intended to solve the problem.

At this stage, however, we leave the world of certainties. Different people arrive at different descriptions of what an outside observer might describe as the same problem with varying computational consequences even if their descriptions are logically equivalent. A variety of primitive constructs can be chosen on which to build the overall structure. When describing family relationships, for example, we can choose to start with a description of parents and their children, with the sex of family members, with an account of fathers and mothers, or with a chain of ancestors. Each can be defined in terms of the others, and different approaches will seem more appropriate in particular social circumstances.

Natural languages such as English are notoriously imprecise and ambiguous, though extremely expressive. 'Words', said Humpty-Dumpty in *Through the Looking Glass*, 'mean what I want them to mean, no more, no less'. Computer scientists have striven towards the use of formal languages such as *predicate logic* that have the virtues of precision, lack of ambiguity, and can be understood and used by both man and machines. The series of research projects described in more detail in the third chapter of this book that use logic as a computer language for children, has indicated against many expectations, that children can make effective use of formal logic as a tool for problem description and solution with the computer, given the appropriate context, structures and representations. Children do not, any more than ordinary adults, want to have to grapple with complex parentheses and quantifiers, with brackets and complex formal symbols. Stripped to its simplest essentials of facts and rules, logic does offer a mediating language between man and machine that can be used at all stages of the process of computational problem-solving. It provides, the 'missing link'.

We still have to give attention to knowledge representation and structures and the ways in which it is natural for humans to deal with problems of knowledge. This has long been the concern of human and social scientists without the aid of the computer. We can offer them a new tool to assist in their work, with considerable benefits to be expected for computer science in general. This book includes descriptions of the results of a number of such attempts in recent years.

## 1.7 Higher-level languages

In his perceptive book *The Computer Comes of Age: the people, the hardware and the software* (MIT Press, 1984), René Moreau describes the development of computer technology to 1963. Although he gives an illuminating account of early computer hardware, he assigns increasing importance to computer software and advanced concepts of computation, and successive generations of higher level computer languages. Our account is centred on the position two decades later, and explores the implications of changing views of computers as summarised by Moreau:

> The early view of a computer was that it was a machine on which users could first develop and then run programs; but it soon became clear that the combination of computer and program could be an extremely powerful tool in the hands of users who need know nothing about computer science but who want simply to process their data with the aid of the program and are content to take on trust from the specialists all details of the method.
>
> Thus two classes of user appeared in the early 1960s. One was the class of computer specialists, a group of whom would share the machine for program development, each effectively unaware of the existence of all the others using the machine at the same time. The other was the class of users who were not computer specialists, a group of whom would share the use of applications programs and data.

Our concern here is with computer users who are not specialists, whose need is to solve particular problems using the computer as a tool, and who are not primarily interested in the computer itself. There has been, and will continue to be, a problem of communication with the machine. Moreau's approach is instructive:

> The user communicates with the machine by means of a programming language. These are of two main types: those intended for systems programming, and therefore used for writing the software that controls the basic resources of the machine; and those intended for the solution of problems on the machine and called problem-oriented languages. From the abstract point of view, the first type is only a special case of the second.

He draws on the analysis of the French structuralist Fernand De Saussure in his account of computer languages:

> A language sets up a correspondence — a set of concepts to be communicated to the machine, termed 'signified' and 'signifiers' by De Saussure, or here 'sources' and 'targets', respectively. The sources are the statements of operations required, and the targets are the corresponding instructions given to the machine. The set of the targets is said to be the 'volcabulary' of the programming language.

In light of this analysis, which makes no concessions to the detailed construction of conventional computers, we can begin to make sense of the increasing proliferation of higher-level computer languages. We should not make the mistake of thinking that a computer language will ever have the flexibility and expressiveness of natural language, but we would normally wish to use a language that allows us to address our problem without having to be an expert on the internal design of the computer.

At the lowest level we can use machine code, a language which consists of instructions which the computer can execute directly, reflecting the internal wiring of the machine. These simple languages are expressed in binary (zeros and ones). Mistakes are easily made because the user needs to know the binary codes for each item in the computer memory. Computer use could not have developed had this remained the only kind of programming language available.

In order to bridge the gap between the language used by the machine and that used by humans to express the method of solution of a problem, two courses of action were available. Problems could be adapted to fit the constraints imposed by the machine, or the communications language could be adapted to the needs of the problems.

As a half-way step a translator program could be developed to offer a high-level language to the computer programmer. The translator would transform a program written in this language into one in machine code understood by the machine. Various translators have been developed, offering a variety of medium and high-level languages.

High-level languages usually fall into two categories: those evolving from trial and error, and those resulting from a prior process of abstract definition. The most recent group of programming languages based on mathematical logic, the functional and logic programming languages, can be seen as progressively uniting the two earlier categories as we learn how to make practical use of powerful abstract notations. Until recently high-level languages were based on the machine architecture defined by von Neumann, with a simple central processor executing a sequence of instructions.

FORTRAN (FORmula TRANslation) was developed by John Backus and colleagues at IBM, with the objective of enabling a problem to be stated concisely in mathematical notation. By 1955 FORTRAN was as efficient as assemblers, the low-level languages used for representing machine-code programs. This made it possible to address the cost of programming, which was already exceeding the cost of computer hardware. It also helped programmers to escape from the limitations of a particular computer by the provision of FORTRAN dialects for most computers. This did not mean that the same programs could necessarily be run on computers of different types, but it did increase the versatility of the users and their expensive skills.

Once it was known that high-level languages were practically possible, more attention was given to their formal definition, emphasising their capabilities for expressing the concepts required in problem-solving rather than the particular features of the machine on which they would be run.

An early attempt (part European, part American) to get away from the notion that only specialised programming languages could meet the needs of users led to the implementation of ALGOL. ALGOL (ALGOrithmic Language) was based on a theoretical model which, it was intended in 1958, could be used 'not only for communication between man and machine but also for communication between people'. Little concern was shown for ease of input and output, and although many of its new concepts, such as block structures in programs, were highly influential in the development of programming languages, ALGOL itself had only limited commercial success.

PASCAL, pioneered by Nikolaus Wirth and Tony Hoare in the 1960s, built on the technical advances of ALGOL, and emphasised good habits of programming which apply whichever high-level language is then used. It has been highly influential in the training and education of computer scientists, though it cannot compete in performance terms with lower-level and less structured languages.

COBOL (COmmon Business-Oriented Language) originated in meetings financed by the American Department of Defense in 1959, where the concern was to develop a portable language with a precise definition which could provide a sound basis for military and civil data processing. Its supporters claim it offers a quasi-natural language for communication between man and machine, and it remains the dominant language in commercial data processing, though it is little taught in academic departments of computer science.

Other formally defined languages have made technical advances, and have been hailed by their exponents as steps towards a universal or multi-purpose language — the Holy Grail of computing. PL/1 (Programming Language 1) has been widely used in a great variety of applications, combining the advantages of FORTRAN and COBOL with the structures of ALGOL, but has still been restricted by the differences between compilers on different machines.

Educational users of computing are likely to have encountered BASIC, designed as an introductory language at Dartmouth College in New Hampshire in the mid-1960s. It became available at the same time as time-sharing systems, which greatly expanded the potential scope for computer use. It was now possible for several users, perhaps many hundreds, to use one computer at the same time, 'sharing' the use of the Central Processing Unit. At a stroke this greatly increased the number of potential programmers, and the economic viability of expenditure on computer hardware. Computing was no longer restricted to specialists, and BASIC provided the introduction to computing for this new clientele.

As microcomputers appeared in the 1970s they were intended for a target market of novices for whom BASIC had also been designed. With BASIC it is easy to write a simple program within minutes of starting to use the language, but by the same token it is almost impossible to develop programs of great scale and complexity which also remain comprehensible to other programmers. With its simple syntax and limited constructs it is easy to implement, albeit in diverse dialects, on most computers, and has helped to sell millions of computer systems, many of which will not have been useful in practical terms. It encourages the writing of programs interactively at the computer keyboard, at the expense of considerations of program and data structure, and can be argued to have a damaging effect on later programming practice.

Social scientists concerned with the language and culture of primitive societies in the fields of social anthropology and linguistics, like Edward Sapir, have identified the limited effects that language can have on perceptions of the world, and BASIC has certainly served to distort the perceptions of new information technology for a whole generation of students. Edsgar Dijkstra, one of the founders of the discipline of structured programming and then of software engineering, has argued vigorously that the use of crude computing constructs (such as the GOTO statement in unstructured forms of BASIC), has highly damaging consequences both in terms of the resulting programs and the disorderly undisciplined ways of thinking that are encouraged. As was agreed by the Alvey Committee, drawing up the United Kingdom's response to the Japanese invitation to participate in their Fifth Generation Computer Project, it cannot be wise to offer today's technology for tomorrow's problem-solvers with yesterday's programming concepts.

Other languages have appeared since 1978, largely for use with microcomputers, which have been intended to meet some of the limitations of the all-pervasive BASIC. COMAL (COMmon Algorithmic Language), developed in Denmark, and LSE (Langage de programmation Structure pour l'Education), developed in France, have fervent followers but have little practical future. They have borrowed features from PASCAL, and can provide an introductory tutorial function, but do not offer a route into the development of large commercial systems.

Since the 1950s there has been a development of functional

languages, with a more formal basis, which have been at the same time easy for advanced specialists and difficult for non-specialists. John Backus has continued his work as a pioneer of programming languages, and spoke in 1978 in his Turing Award Lecture to the Association for Computing Machinery, of the problem of the 'von Neumann bottleneck' produced by the use of a single central processor. He said that the programmer loses sight at times of the fundamental nature of the process being used for the solution of a problem and becomes preoccupied with the problems of managing the traffic flow through the bottleneck. Backus argued for the separation of instructions and objects in a program, giving rise to novel approaches to computation.

LISP (LISt Processing) was developed in 1958 by John McCarthy, for work in artificial intelligence research, with a particular emphasis on list processing. In its purer forms it can be contrasted with 'von Neumann languages', in that it has no concept of instruction and the same structure is used for both a LISP program and for data. It was for twenty years the dominant language in American research laboratories concerned with natural language processing, robotics, and Expert Systems, with which we will deal in later chapters. Until recently it has required the use of mainframe or minicomputers, needing too much computer memory to be implemented on low-cost microcomputers.

LOGO (derived from the Greek 'logos', or 'knowledge') was developed by Seymour Papert and colleagues at Massachusetts Institute of Technology in 1967, originally as a dialect of LISP but with a particular emphasis on education use. Typically introduced to children and students through 'turtle graphics', it is a powerful tool for introducing concepts of procedural thinking and programming, as well as a potentially powerful programming language in its own right. It has undergone a resurgence of popularity as personal and educational microcomputers have been provided with sufficient memory to support implementations of LOGO, but has failed to lead, as was promised, to intelligent uses of computers across the curriculum and in diverse areas of application. In a later chapter we examine further its use in education.

SMALLTALK resulted from many years of research at the Xerox Palo Alto Research Centre led by Adele Goldberg. Although it had elements in common with the object-oriented simulation language SIMULA (developed in 1962) it had few practical applications until recently, largely because of the great demands it placed on computer memory. Its influence can be seen in recent microcomputer systems such as the Apple LISA and Macintosh, and it is explored further in chapter 3.

PROLOG (PROgramming in LOGic) was first implemented in 1972 in Marseilles by Alain Colmerauer and colleagues, but remained a tool for artificial intelligence researchers rather than for commercial applications until the Japanese plans for fifth generation computer systems were announced in 1981. It was the first computer language to give practical realisation to Robert Kowalski's theoretical ideas of logic programming.

Kowalski's theoretical advance was the separation of the logic of a program from its control features, coupled with the discovery that both could be expressed in the same formal notation — a type of logic called *predicate logic*. This provided a crucial connection between previously separate aspects of computer science, all of them making use of predicate logic. These included, databases, natural language processing, systems design and specification, robotics, artificial intelligence and programming theory. Recent research programmes have succeeded in developing new practical systems which begin to embody Kowalski's theoretical insights.

There are those who will argue the case for one or other computer language as being universally applicable, but all such arguments are fundamentally unconvincing. Forty years of research and development have taught us that there are no objective criteria for the form of language that a user should choose when wishing to communicate with a machine. As Moreau concludes:

> Programmers will doubtless prefer a language specifically adapted to the needs of a well-defined field if they need to solve problems in that field, and a more 'universal' language if they are concerned with problems from several fields.

It is part of our thesis that computer use should not be equated with computer programming in the conventional sense. Any means that are used to communicate with a computer can be seen as constituting a computer language, although often the communication is not intended to take the form of issuing a sequence of instructions. We could instead be using a language to describe information to the computer, to ask or answer questions, or to communicate with other human beings past, present or future.

## 1.8 User-friendly systems

In recent years computer manufacturers have increasingly made the claim that their products are user-friendly. This has possibly been to distinguish them from the offerings of their rivals which are to be regarded as user-hostile. Each of these descriptions depends on the flawed assumption that there is an understanding of the user; that is to say, the needs and the difficulties that the user faces in interacting with computer technology. These issues have been the concern of researchers in what can be called 'man-machine interface', human-computer interaction', and more generally 'cognitive science'. Common to each of these areas is the study of a field created by the very existence of the technology, and the assumption that in some sense we interact with computers just as we interact with other human beings.

Brian Gaines and Mildred Shaw have addressed these issues in their book *The Art of Computer Conversation* (Prentice-Hall 1984), where they recognise the new phenomenon to be addressed:

> The art of conversation, of making oneself clear to the other person, of using a technology he will understand, of expressing information in alternative ways, of interpreting what the other person says, all these were unnecessary in the highly technical worlds of the early computers. In any event computers were the very expensive and godlike creatures with whom it was appropriate for their servants to converse in a formal and stylised fashion. The courts of Kings and Emperors have always been so. The art of conversation with computers did not die. It has not yet developed.

Gaines and Shaw place great emphasis on what they call 'dialogue engineering', building on the work of researchers such as Gordon Pask at Brunel University who developed 'conversation theory' in 1965 in the context of work in cybernetics:

> Dialogue engineering is a new technology at the heart of personal computing and interacts strongly with all the other computing technologies. The acquisition and presentation of numerical data, the capability of the operating system to be presented and controlled simply and logically, the capability of the database to accept information in a form natural to people without distortion — all of these involve the psychology of people as much as they do the technology of the computer.

Unusually among specialists in computer technology, they are able to see their technical concerns in a human and social context, which serves to illuminate by reflection some of the technical issues to which we have to turn:

> Whatever attitudes we adopt to computer technology, it, like other products of the human imagination, now has an existence of its own that influences us and is not wholly under our control. We created computers but now computers are coming to create us, to change our lifestyles, our environment, our jobs, our leisure activities, our freedom of expression, our knowledge of ourselves and of others, and their knowledge of us. Humankind is highly adaptable and as we change the nature of our world we change with it. Our media are extensions of ourselves and part of our evolution and the growing symbiosis of people and computers may be a major step in the evolution of the human race.

It is indicative of the promise of user-friendly systems that computers can catalyse such musings. Clearly matters have progressed from the batch processing of punched cards or tape, with a considerable time lag between input by the user and output from the machine. It is analogous to the difference between correspondence with a friend or business associate in another country, and a face-to-face conversation. Time is an important factor, but far from the only one. Studies have shown that personal productivity of programmers is closely related to the speed of response of the computer system, and this is of course one of the driving forces behind the continued pressure for greater performance of hardware and efficiency of software.

In recent years there has been a decline in the significance of large centralised computer centres, and a greater 'distribution' of computer processing power and control. Large mainframe installations with rows of glowing vacuum valves have been succeeded by smaller minicomputer systems, with first transistors, then progressively larger scales of integrated circuitry, and finally small microcomputers in the hands of non-specialist individuals whose computing power is greater than that of their larger predecessors.

'Am I my computer's user, or is it using me?' This has been a frequent question from critical observers of the social aspects of computers. Seymour Papert, in his book *Mindstorms* (Basic Books and Harvester, 1980) distinguished between programming a computer and being programmed by it. He recalled his childhood fascination with gears and other parts of cars, and associated this with the developmental psychology of Jean Piaget:

> Piaget's work gave me a new framework for looking at the gears of my childhood. The gear can be used to illustrate many powerful 'advanced' mathematical ideas, such as groups or relative motion. But it does more than this. As well as connecting with the formal knowledge of mathematics, it also connects with the 'body knowledge', the sensori-motor schemata of a child. You can be the gear, you can understand how it turns by projecting yourself into its place and turning with it. It is this double relationship — both abstract and sensory — that gives the gear the power to carry powerful mathematics into the mind.

Papert summarised his thesis:

> What the gears cannot do the computer might. The computer is the Proteus of machines. Its essence is its universality, its power to simulate. Because it can take on a thousand forms and can serve a thousand functions, it can appeal to a thousand tastes.

One of the authors of this book, Richard Ennals, gave an analogous account of his approach to computing derived from history teaching, in an article in 1979 in *Practical Computing:*

> The logical structure of a computer analogue of a historical situation is similar to the equivalent classroom simulation, with the possible difference

that the simulation game normally involves decisions by a number of participants rather than by a single player. So it should be feasible to make use of the computer analogue to enhance the classroom simulation, and vice versa.

There is philosophically a clear link between mathematical and computer logic, language structure and the rules of games — as outlined, for instance in Wittgenstein's *Philosophical Investigations* and Richard Braithwaites's *Theory of Games as a Tool for the Moral Philosopher*. This has clear, if undeveloped, practical implications, especially for people like teachers who have the task of explaining complex processes.

What kinds of user interaction have been available? In physical terms there has been a progression from batch processing to interactive computing. For some years this took the visible form of the 'glass teletype', as if a conversation was being conducted between typewriters and displayed on a screen. The user typed in a sentence in an artificial lnaguage and received a typed response.

There is a continued mystique surrounding the output of computers, which is often given a status exalted above the output of mere people. The phenomenon of the cargo cult of the South Pacific Islands, where prayer is made to models of radio sets in the hopes that aeroplanes will arrive with supplies of American consumer goods as in the days of the Second World War, is reappearing in a revised form with computers. They are somehow seen as offering the key to knowledge and economic success. People still tend to place reliance on the computer prediction of an election result, forgetting that it will simply be the consequence of the application of formulae provided by a human designer. People who believe the predictions of such systems are truly being used by them. When people are faced with enormous gas bills prepared by a computer, they may tend to pay rather than complain as they would to a human clerk. We should ask the questions: if we do not insist on participating in the solution we will just be incorporated into the problem.

Brian Gaines and Mildred Shaw in *The Art of Computer Conversation* have made a noble attempt at classifying some of these issues, putting forward 30 'proverbs' to guide computer conversation. We quote some of their proverbs as illustrations of the wealth of intriguing questions to be explored.

We are all responsible for computer behaviour.
Computers provide a new medium for communication.
Users already have expectations about computers.
Users readily think of computer systems in the same way they think of people.
The style of conversation varies with the technology used.
The system should adapt to the user under user control.

Programs create the reality experienced by the users of computers.
Fluent language may not imply fluent understanding.
Design for a changing and uncertain future.
Make the best use of today's technology today and tomorrow's tomorrow.

**Figure 1** Chronology of early computing technology

---

1801 Joseph-Marie Jacquard in France develops a loom programmer by punched cards.

1823 Charles Babbage in England plans a mechanical machine able to 'solve any equation and to perform the most complex operation of mathematical analysis'.

1890 Herman Hollerith's punched cards and tabulating equipment are used in US Census.

1893 Otto Steiger in Switzerland invents the first efficient four-function calculator, a mechanical device known as the Millionaire.

1939 At Bell Labs, George Stibitz builds with telephone relays the first binary calculator.

1940sElectromechanical and electronic calculators first make use of vacuum tubes; these components lead to the evolution of the von Neumann electronic computer in the United States. In Britain similar advances lead to the Manchester Automatic Digital Machine, which in 1948 first employs a stored program.

---

Although not all of the inventions of early computing technology are shown here, the main ones are. Especially important in shaping the electronic computer was the technology available in the early 1940s. Stibitz's calculator was built on the binary principle because of the 'all-or-nothing' property of telephone relays. In the same way the single central processor of the later von Neumann computer was also restricted to machine code, a programming language dependent on binary numbers.

**Figure 2** Five generations of computer technology

| Generation | Characteristic electronic component | Advantages | Disadvantages |
|---|---|---|---|
| 1st generation 1940-1952 | vacuum tubes | vacuum tubes were the only electronic components available | large size and plenty of power required generated heat, so air-conditioning and constant maintenance were necessary |
| 2nd generation 1952-1964 | transistors | smaller size, less heat generated, more reliable, faster | needed air-conditioning and maintenance |
| 3rd generation 1964-1971 | integrated circuits | even smaller, even lower heat, less power needed, even more reliable, faster still | initially, problems with manufacture |
| 4th generation 1971- | large-scale integrated circuits | no air-conditioning,minimal maintenance, high component-density, cheap | initially less powerful than earlier mainframe computers |
| 5th generation | | Currently under development | |

This figure shows the five generations in the development of the electronic computer. Early electromechanical and electronic calculating machines built in the 1940s covered large floor spaces and consumed large amounts of power. Size and heat remained a problem until the 1960s, when integrated circuits allowed the construction of the first compact computers. Since then, the microcomputer has shown what can be achieved even with a relatively small machine. Note that there is no hard and fast dividing line between third and fourth generation computers.

**Figure 3** Chronology of computing concepts

| time | technology | concepts |
|---|---|---|
| 19th century | mechanical wheels, cylinders (unsuccessful) | mathematical analysis |
| 1940s | von Neumann electronic computer — cumbersome vacuum valves | machine languages (based on binary numbers) |
| 1950s | more reliable computers — from 1957 transistorised | first high-level programming languages, FORTRAN; numerical computation |
| 1960s | smaller, faster computers — from mid-1960s electronic chips | program and data structures; ALGOL, PASCAL; operating systems, time-sharing; object-oriented programming, SIMULA |
| 1970s | large-scale integrated circuits | software engineering; data abstraction; logic programming, PROLOG |
| 1980s | | interface technology: hardware—displays, windows, graphics software — modules, specifications, software interfaces; ADA |
| 1990s | very large-scale integrated circuits (VLSI) | knowledge engineering; parallel, interactive, automatic programming; natural language, logic, expert systems |

# 2 The Fifth Generation

## 2.1 Computing in crisis

Computing is encountering a crisis of expectations. Individuals and companies have purchased computer systems and have been led to believe that by this step they have also purchased the solution to their problems. In some cases the reverse has been true, and the purchase of a new system has been closely followed by the financial collapse of the purchaser. Computer systems tend to magnify any deficiencies in the underlying system of management used in a company.

Off-the-shelf packages, whether database systems or integrated solutions, will probably be based on successful use by the original designer in the original real-life setting. The situation of the user company is always different, and unless the staff in the user company have a detailed understanding both of their problems and of the workings of the computer system, difficulties can be expected. This is rarely made clear when systems are sold, and neither accompanying documentation nor the brief introductory course which may accompany the introduction of the system will normally make this clear.

Purchasers with real-life problems have been led to believe that computer systems are easy to use, reliable, and lead to greater efficiency and productivity on the part of the user. None of these artificially-induced expectations has been borne out by experience in more than a small proportion of cases, where the problems addressed were well described and the performance of the system matched the underlying requirements of the user. A custom-made database or payroll package may suit its original customer, but may prove inappropriate in different circumstances.

Automation and high-technology production methods have been successfully applied to aspects of computer hardware. Computer systems are used to design other computer systems, and processing power that used to require a large system with an air-conditioned room can now be printed on a single chip of silicon or gallium arsenide. The scale of problems and of the software needed to solve them is not so easily reduced. The field is insufficiently understood and matters are not improved by the prevailing expectations of instant solutions.

The current software crisis is such that the major component of the cost of computer systems is now that of software development and maintenance. Data processing departments have vast backlogs of work leading to frustration and disillusionment. Change is made doubly difficult by the fact that departments soon develop a commitment to their current computer systems, which are often incompatible at the software level with supposedly superior alternatives. A great deal of time may have been devoted to designing, implementing and entering data for a company database system, there may be pressures to resist a change in hardware or software approaches that might mean the abandonment of such work. This has not until recently bothered computer manufacturers unduly, but helps to explain the current emphasis on networks communications and system standards. Using a computer network, independent machines can be connected and share programs and data, as long as there are no incompatibilities in their design. There is growing pressure for the availability of network systems and communications that permit machines made by different manufacturers to communicate. Without the offer of a smooth transition and updating of existing systems, customers will resist further purchases and the industry will grind to a halt. This point has also to be noted by the developers of new generations of computer systems, which must allow users to continue to use previous systems which are still performing well.

One of the products of this crisis is the new discipline of software engineering, whose objective is to develop a rigorous engineering approach to software design, implementation and maintenance. While many other science disciplines from bridge-building to automobile manufacture are conducted within known engineering principles, it is only now being recognised that computer programming must be viewed in the same light. A computer program in the future must be less a poem and more like a reasoned argument. Software engineering involves modular design, competent testing, and rigorous diagnostics. Without all these changes computer programmers cannot expect their efforts to be trusted and understood by the general user.

Despite the advent of better software design principles the crisis of expectations will continue. What is needed is a fundamental shift in the type of applications to which computers are put.

## 2.2 Machines thinking like people

### Some simple knowledge-based systems

We have already made mentioned in passing expert systems. An expert system, also known as an 'intelligent knowledge-based system,' is a computer program that in some senses mimics the activities of a human expert. Before meeting expert systems it is important to understand the basic ideas and mechanisms from which they are constructed.

The following examples are written using MITSI (Man In The Street Interface), which was developed in 1983 by Jonathan Briggs to explain ideas of logic programming and knowledge-based computing to non-specialists. These programs concentrate on describing the problem, rather than on how it is to be solved. One consequence of this is that the same programs will run, with minimal changes, on the new generation of parallel computers.

## A small expert system for bicycle repairs

An expert system will have a 'knowledge base' containing knowledge about the subject area to be covered by the system. Using MITSI we can describe knowledge about bicycles, starting with the following facts:

```
puncture causes flat
leaky-valve causes flat
flat causes uneven-ride
broken-spoke causes distorted-wheel
distorted-wheel causes uneven-ride
distorted-wheel causes erratic-braking
faulty-cable causes erratic-braking
erratic-braking causes accident
```

An expert system will also have general rules to enable the system and the user to manipulate the facts and perform deductions. We can add, for instance, two simple rules to our program, describing bicycles as systems in which one fault can lead to another:

```
FAULT leads-to PROBLEM
  if FAULT causes PROBLEM

FAULT leads-to PROBLEM
  If FAULT causes GLITCH and GLITCH leads-to
  PROBLEM
```

An expert system will be capable of interrogation by the user in a straightforward manner. We can ask simple questions and receive answers:

```
puncture causes flat?
YES

WHAT causes flat?
  puncture causes flat
  leaky-valve causes flat
no (more) answers

WHAT leads-to uneven-ride?
```

```
    flat leads-to uneven-ride
    distorted-wheel leads-to uneven-ride
    puncture leads-to uneven-ride
    leaky-valve leads-to uneven-ride
    broken-spoke leads-to uneven-ride
  no (more) answers
```

An expert system will be able to explain its reasoning to the user, rather than simply presenting an answer.

In MITSI we can obtain explanations of a conclusion in terms of the immediate conditions which had to be satisfied, using the same rules as above in reverse. This is very much a minimal approach to explanation, designed for use on small microcomputers. Real-world systems will require much greater sophistication.

```
Why broken-spoke leads-to uneven-ride?
broken-spoke leads-to uneven-ride because
  broken-spoke causes distorted-wheel and
  distorted-wheel leads-to uneven-ride

Why distorted-wheel leads-to uneven-ride?
distorted-wheel leads-to uneven-ride
  because
  distorted-wheel causes uneven-ride

why distorted-wheel causes uneven-ride?
distorted-wheel causes uneven-ride is
  stated
```

We would like our expert system to be flexible in its use, permitting a variety of questions to be posed without the necessity of preprogramming, or of predetermining the manner of use. As well as reasoning back from an accident or an uneven ride to a broken spoke, we can reason forward and explore the consequences for the system of a broken spoke.

Note that each of the answers to the questions is generated by a process of deduction, and does not have to be entered explicitly. This is one of the advantages of using a logic database made up of facts and rules, rather than the less powerful relational database approach. Thus:

```
broken-spoke leads-to WHAT
  broken-spoke leads-to distorted-wheel
  broken-spoke leads-to uneven-ride
  broken-spoke leads-to erratic-braking
```

```
   broken-spoke leads-to accident
  no (more) answers
```

we might add a final advisory rule:

```
FAULT should-be fixed
  if FAULT leads-to accident
```

and a final question:

```
WHAT should-be fixed?
  erratic-braking should-be fixed
  distorted-wheel should-be fixed
  faulty-cable should-be fixed
  broken-spoke should-be fixed
no (more) answers
```

Interestingly, though a puncture and leaky valve both lead to an uneven ride, our program has not established a causal connection from either of these faults to an accident, so we are not at present advised to fix them. We can, if we wish, remedy this by adding a further explicit fact, such as

```
flat causes erratic-braking
```

## Some well known knowledge-based systems

Research and development has been conducted in the United States for 25 years, and some well known systems have been developed, though typically they have not been widely used for the performance of practical tasks. The following examples have been chosen to illustrate how these famous systems, notwithstanding the expensive equipment and many years involved in their development, are essentially similar in form to our simple examples.

### *MYCIN*

MYCIN was developed at Stanford University in 1965 by a team led by Edward Shortliffe, in association with the Medical School, and was concerned to represent the knowledge and replicate the procedures of expert medical diagnosis in a particular defined area, concerned with the administration of antibiotics. The original system was built using LISP, for 20 years the dominant language for artificial intelligence research in the United States. It was based on the principle of production rules, which take the form:

```
IF condition and condition
THEN conclusion
```

A typical rule follows:

```
IF
  1  the site of the culture is blood and
  2  the identity of the organism is not
     known with certainty, and
  3  the stain of the organism is
     gramneg, and
  4  The morphology of the organism is
     rod, and
  5  The patient has been seriously
     burned.
THEN
  There is a weakly suggestive evidence
  (0.4) that the identity of the organism
  is pseudonomas
```

It will be clear from the above that the formulation of the rules depends on the expert medical knowledge of members of the expert systems team, which will typically include subject domain experts and knowledge engineers, whose job is to elicit the knowledge from the expert and represent it in a form which the computer can understand. It is easiest to work in a subject area where the classification of knowledge is generally agreed and where procedures can be described in clear explicit terms. The contribution of the computer is to cope with the volume and complexity of information that would be beyond the unaided user. You will note that the conclusion to this rule includes some provision for uncertainty, and there are often situations where more than one answer could apply, and different weights can be assigned to each.

The MYCIN system has not been widely used by doctors, whose patients have mixed feelings about computer diagnosis. It has, however, provided a classic example with which researchers can try out their ideas. From MYCIN was developed EMYCIN, which preserved the inference mechanism, dialogue system and approach to explanation but removed the medical information, leaving a so-called 'expert system shell' into which new knowledge of an appropriate form could be put. EMYCIN has in turn been reimplemented in other computer languages such as PROLOG and POP-2 (developed in Edinburgh in 1974 by Robin Popplestone and colleagues), as a means of testing their capabilities.

A further application of MYCIN, which will be discussed in more detail later, was as an expert system to accompany an intelligent tutoring system, teaching skills of medical diagnosis without having to make use of real human patients. The GUIDON system, developed by Clancey in 1973 and building on the basis of work in MYCIN, was intended to lead the learning doctor through a diagnostic session, helping him to understand the problem, his mistakes, and the system's view of the problem. Again you will note from the following rule from the GUIDON system that the same production rule approach is maintained:

```
IF
  1 There are rules which have a bearing
    on the goal which have not succeeded
    but have not been discussed
  2 The number of rules which have a
    bearing on this goal which have
    succeeded is 1
  3 There is strong evidence that the
    student has applied this rule
THEN
Simply state the rule and its conclusion
```

*PROSPECTOR*

PROSPECTOR was developed at Stanford Research Institute in 1978 by a team led by Richard Duda to assist in mineral exploration. Given a mass of geological data concerning a region, and heuristic rules elicited from experienced prospectors, the intention was to identify areas that offered the best prospects for the discovery of valuable minerals. Unsuccessful drilling operations cost millions of dollars, so considerable financial benefits were available, with the costs of the research and development seen as small by comparison. The PROSPECTOR system entered high-technology mythology when it assisted in the identification of previously unknown deposits of molybdenum. You will note that the system is again based on production rules, such as:

```
IF
  There is hornblende pervasively
  altered to biotite
THEN
  There is strong evidence (320, 0.001)
  for potassic zone alteration.
```

Such a complex system is economically viable in areas where expertise is scarce and the economic benefits are great. Similar systems have been developed to assist in oil exploration, such as the Schlumberger Dipmeter-Adviser.

*R1 or XCON*

The R1 (known commercially as XCON) system was developed at Carnegie-Mellon University by John McDermott and colleagues in 1979 to assist in the task of configuring computer systems, which are made up of different combinations of components depending on the needs of the customer. The work was done in association with Digital Equipment Corporation, the largest American manufacturers of minicomputers, who subsequently implemented the system as XCON. It was implemented in the language OPS5, an experimental language for building expert systems, again based on production rules, and itself usually implemented in LISP. An example rule from R1 follows:

```
IF
  1  the most current active context is
     selecting a box and a module to put
     into it, and
  2  the next module in the optimal
     sequence is known, and
  3  the number of system units of space
     that the module requires is known,
     and
  4  that box does not contain more
     modules than some other box on a
     different unibus
THEN
Try to put that module in the box
```

The system was extremely expensive to develop, and has undergone many changes. It was regarded as a justifiable expense as it helped to motivate research and development in expert systems at Digital, whose superminicomputer hardware is intended for use in artificial intelligence and other advanced applications. Similar systems are now in use in most major computer hardware companies.

In each of the above cases the rules have the attraction of being comprehensible; experts can recognise their knowlege in the program, and the program follows such rules, citing them in its explanations.

*AM*

AM is a system concerned with the formation of mathematical concepts, and was developed by Douglas Lenat of Stanford University in 1979. He took the idea of rules further, as he tried to explore the complex area of mathematical concepts. His hope was that by representing concepts in facts and rules, and giving them to the expert system, the system might be able to explore the area and uncover new rules and concepts previously undiscovered or not normally stated explicitly. The following two rules used by AM are applied to concepts, examining them in search of interesting cases:

```
IF
  Concept C possesses some very interesting
  property lacked by one of its
  specialisations S,
```

```
THEN
  Both C and S become slightly more
  interesting
IF
  There are no known examples for the
  interesting concept X
THEN
  consider spending some time looking for
  such examples
```

*Using rules*

We can see that there is nothing new in using rules as such. There follow some traditional rules found in a manual for operators of lime kilns:

1 When the burning zone temperature is drastically low
   (a) reduce kiln speed
   (b) reduce fuel

2 When the burning zone temperature is slightly low
   (a) increase I.D. fan speed
   (b) increase fuel rate

What is new is that rules which make sense to us can also make sense to the computer.

## Conversations with knowledge-based systems

There are those who hold the mistaken view that it is possible to communicate with a computer in ordinary language. It is not possible now, and although research and development work is producing systems that come ever closer to the goal, we can expect there to be continuing restrictions on our use of vocabulary, complex sentence structures, and complex ideas.

There have been a number of significant projects involving the use of natural language with computers, with most of the work until recently being conducted in the United States. It is still a subject that has raised little interest among British companies, possibly because they assume machines, like all foreigners, will be obliged to speak English.

*ELIZA*

Joseph Weizenbaum of Massachusetts Institute of Technology was concerned to demonstrate the limitations of the intelligence of computer technology when, in 1965, he developed his famous ELIZA system (named after Eliza Dolittle in Bernard Shaw's *Pygmalion* and the musical *My fair Lady*). A particular program using ELIZA simulated the behaviour of a non-directive psychotherapist of the school of Carl Rogers, in conversation with a patient. His system was based on the simple idea of recognising patterns in the sentences from the patient, and responding mechanically with pre-set adapted responses. There follows a dialogue between a user and ELIZA:

```
USER:  Men are all alike.
ELIZA: In what way?
USER:  theyre always bugging us about some-
       thing or other.
ELIZA: Can you think of a specific example?
USER:  Well, my boyfriend made me come here.
ELIZA: Your boyfriend made you come here.
USER:  He says Im depressed much of the time.
ELIZA: Im sorry to hear that you are
       depressed.
USER:  Its true. I am unhappy.
ELIZA: Do you think coming here will help you
       not to be unhappy?
USER:  I need some help, that much seems
       certain.
ELIZA: What would it mean to you if you got
       some help?
```

To Weizenbaum's intense frustration his program provoked the response he least wanted. People were impressed with the 'intelligence' of the system, and even his secretary preferred to confide in ELIZA rather than to talk to him. This mistaken popular perception of natural language understanding by machines was strengthened by the performance of the computer HAL in the highly successful film *2001: A Space Odyssey*. Science fiction has far outstripped science fact.

*PARRY*

PARRY was built in 1973 by Kenneth Colby of Yale University, who was interested in abnormal behaviour, and built his system to simulate a paranoid patient in conversation with a doctor. A portion of dialogue follows:

```
Doctor:  How do you like the hospital?
Patient: Its not helping me to be here.
Doctor:  What brought you to the hospital?
Patient: I am very upset and nervous
Doctor:  What tends to make you nervous?
Patient: Just being around people.
Doctor:  Anyone in particular?
```

It will be clear that one of the interests of artificial intelligence researchers is the simulation of behaviour. Mental illness is an obvious area of interest, because the disorders of patients are described in terms of stereotypical patterns of behaviour, which can be modelled in computer programs in order to be subjected to scrunity and compared with observations of real patients.

Another case of abnormal behaviour modelled by Colby and his colleague Abelson at Yale was the Cold War Warrior. Their 'ideology machine', based on a close study of the policy and speeches of Senator Barry Goldwater and his supporter Ronald Reagan, tried to demonstrate a simple black and white view of the world, tested out in a question and answer session such as might be held with a spokesman for the White House.

It is much harder to conceive of a system that could deal with the broad range of human responses, so conversational systems tend to focus on a particular need or situation.

### *SHRDLU*

SHRDLU is a highly influential system developed in 1973 by Terry Winograd of Massachusetts Institute of Technology, who was interested in using a substantial part of natural language when dealing with objects and events in a very restricted area. He devised a 'blocks world', reminiscent of the opening sections of Wittgenstein's *Philosophical Investigations*, within which commands could be given to a robot to pick up objects of different shapes and colours, in order that it could carry out certain tasks like building a tower or placing objects in boxes. A portion of dialogue follows, which indicates the flexibility and richness of the converstion that was possible:

```
USER:    Pick up a big red block.
SHRDLU:  OK.
USER:    Grasp the pyramid.
SHRDLU:  I dont understand what pyramid you
         mean.
USER:    What does the box contain?
SHRDLU:  The blue pyramid and the blue block.
USER:    What is the pyramid supported by?
SHRDLU:  The box.
USER:    How many blocks are in the box?
SHRDLU:  Four of them.
```

The level of sophistication of the system, in terms of its capacity to deal with complex sentence structures and dialogue, was enormously in advance of ELIZA,but again misled some observers as to the degree of intelligence which could be expected from a computer system.

### *INTELLECT*

By contrast, INTELLECT is a commercial system, developed by Intellicorp of California in 1979 and used to provide an interface to commercial databases, saving the user from having to be expert in the obscure query languages and relational algebras. It is providing an introduction for numerous companies, particularly in the United States, to the potential applications of research and development in natural language understanding. Groups at Stanford Research Institute and in many other research laboratories are now developing new systems with further degrees of sophistication.

The following sample interactions show what the user types, how it is modified by the INTELLECT program, and the response from the database.

```
GIVE ME THE NAMES OF ALL EMPLOYEES WHO HAVE
JOBS WORKING AS A SECRETARY IN THE CITY OF
CHICAGO
PRINT THE NAME OF ANY EMPLOYEE WITH CITY =
CHICAGO AND JOB = SECRETARY
THE NUMBER OF RECORDS TO RETRIEVE IS 30
  SMITH
  JONES
  BROWN
PRINT THE NAMES OF ALL THE CHICAGO EMPLOYEES
WORKING AS SECRETARIES
PRINT THE NAME OF ANY EMPLOYEE WITH CITY =
CHICAGO AND JOB = SECRETARY
PRINT THE SALARY OF BROWN AND JONES
PRINT THE SALARY, AND NAME OF ANY EMPLOYEE
WITH NAME = BROWN OR JONES
THE NUMBER OF RECORDS TO RETRIEVE IS 269
  $22,500      JONES
  $23,250      BROWN
```

### *ASK*

Natural language interfaces to databases and other systems are also becoming available for microcomputers, sometimes implemented in more traditional languages such as PASCAL. One example is ASK, available since 1983 from Hewlett-Packard. The sample interaction shows the questions from a user and the responses from the system.

```
What cities are the home ports of ships
whose
destination is London?
```

```
    Boston
    New York
    London
    Norfolk

  Are there ships that do not have a cargo?
    Yes

  What is the number of New York ships?
    There are two answers
    1)  New York (home port) ships
        1
    2)  New York (destination) ships
        2

  What ships whose cargo is wheat have London
  or Oslo as destination?
    Maru
    Alamo

  What is the cargo and home port of the Maru?
    cargo                home port
    wheat                Tokyo
```

*GUS*

Daniel Bobrow of Stanford University has long experience of research in knowledge representation and natural language understanding. His GUS system, implemented in 1982, is concerned with handling the booking of airline tickets, and involves a significant degree of sophistication. For example, the system recognises what is meant when pronouns are used. GUS represents an attempt to build a system which could attempt the 'Turing Test', whereby the user could have a real-life dialogue with the system without realising that his problem was being solved by a computer rather than by a human airline booking clerk.

```
GUS:    Hello. My name is GUS. I can help you
        plan a simple trip by air.
        Where do you want to go?

Client: I want to go to San Diego on May 28.

GUS:    What time do you want to leave?

Client: I must be in San Diego before 10 am.
```

```
GUS:    Would you like PSA flight 102 that
        arrives at 9.15am?

Client: What is the next flight?

GUS:    Air California flight 310 that
        leaves at 8.30 am and arrives at
        10.00 am.

Client: Ill take the first one.

GUS:    In what name shall I make a
        reservation, please?

Client: Dan Bobrow.

GUS:    I have confirmed the following
        flight:
        PSA flight 102 on Wednesday May 28
        from San Jose to San Diego that
        leaves at 7.30 am and arrives at
        9.15 am. What date do you want to
        return on?

Client: On Friday in the evening.

GUS:    Would you like the flight that leaves
        at 7.45 pm?
Client: Thats fine.

GUS:    I have confirmed the following
        flight:
        PSA flight 307 on Friday May 30 from
        San Diego to San Jose that leaves at
        7.45 pm and arrives at 9.30 pm.
        Thank you for calling.
        Goodbye.
```

*TEIRESIAS*

TEIRESIAS is a development of the work at Stanford University in early 1970s on the DENDRAL system, led by John Gaschnig. This was concerned with mass spectrography and chemical structures. As with the MYCIN system, the attempt was made with TEIRESIAS to apply the approach to reasoning and to holding a dialogue with the user to new domains. In this example TEIRESIAS is being used to advise on the management of investments.

```
Investors name?
** FRED SMITH

Age?
** 40

Present taxable income?
** 40,000

Number of dependants?
** 5

Occupation?
** LAWYER

Amount of previous investment
experience?
(slight = less than one year;
moderate = 2 to 4 years;
extensive = more than 4 years)
** MODERATE

Does the investor subscribe to the Wall
Street Journal, or any
securities market newsletters?
** Y

.........................................

The following investments appear to be
most appropriate at this time:
  Varian-Corporation
  American-Telephone-and-Telegraph

Some of the rules used in this new domain of
investment management are:

RULE 116
If 1) the desired return on the investment
      is greater than 10%,
   2) the time-scale of the investment is
      long term,
   3) the number of dependants of the client
      is less than 3,
```

```
   4) the age (in years) of the client is
        less than or equal to 35
Then there is evidence (.4) that the area of
the investment should be high-technology


RULE 383
If 1) the income-tax bracket of the client
        is 50%,
   2) The client follows the market
        carefully,
   3) the amount of investment experience of
        the client is moderate,
Then there is evidence (.8) that the area of
the investment should be high-technology
```

## 2.3 Problems of knowledge

Although advanced information technology has advanced considerably in the forty years of working computer systems, fundamental problems of knowledge remain unsolved. We should look with new respect and interest at the work of philosophers, and not make the mistake of assuming that philosophical problems can be solved by the application of technology.

The advent of expert systems indeed presents new problems about how we deal with knowledge. The simple examples we demonstrated in section 2.1, for instance about bicycle faults, are contrived and many of the difficulties have been skated over. This is essential, however, when the material is first shown to teachers or students. In later lessons these problems can begin to be explored. Let us now identify some of the problems.

1. We can never have full and complete knowledge of the real world. We have to make do with incomplete information and understanding, and this limitation continues even when we make use of 'intelligent machinery'.
2. We must not confuse the status of our models of real world problems, which can be represented as computer programs, and the real world problems themselves.
3. Knowledge often has to be coded within the constraints of language. Where computer languages are involved this coding may present considerable difficulty.
4. Understanding utterances in natural language presents formidable obstacles to intelligent human beings, and insuperable problems for machines.

5 We must not confuse explanation, which can be clear, consistent and coherent, with prediction, which needs to be complete to be reliable. It does not follow from the fact that we can offer acceptable explanations of past human behaviour that we can provide reliable predictions, determining which of a number of possibilities will turn out to be the case.
6 The knowledge contained within an expert system can only be as complete and correct as that held by the human experts who created it. If computer specialists have manipulated this knowledge, its consistency becomes even more doubtful.
7 The user will interact with the computer expert in a different way to a human expert. Despite the facilities that may explain the answers produced, there is no replacement for the subtle ways in which humans can interact.
8 The computer is unable to provide the user with a guide to the real value of the information provided. Computerisation is guaranteed to increase the quantity of information available but not the quality.

## 2.4 Integration of advanced technologies

Before continuing our exploration of knowledge-based computing, it is necessary to draw into the arena other technologies that will affect the ways in which we manipulate and use information.

Few people are unaware of the increasing effectiveness of photocopying, and the falling cost of photocopiers. The new laser printers, increasingly popular for providing the 'hard copy' printed output from computer systems, employ the same technology as the current generation of photocopiers.

Facsimile transmission has also undergone similar technological advances and price reductions, and also uses common technological components Facsimile transmission enables the user to send the full text of printed or other documents, by electronic means, to any location in the world which is equipped with a low-cost receiver and printer. Desktop publishing is no longer science fiction; the products of computer programs can be printed to professional standards and transmitted around the world in fractions of a second. The implications for the previous mechanisms and institutions devoted to the production, distribution and exchange of information are likely to be revolutionary, in all senses of that word. It no longer requires a large budget to publish and distribute information.

As this sentence is being typed on the keyboard of a word processor it could also be appearing on the screens and printers of colleagues who are connected by electronic networks. Without any necessary cost beyond the original cost of installing receiver equipment, textbooks could be produced for Third World countries, transmitted to remote villages by means of satellite systems which are already in place. On arrival the text could be saved on disk for later reproduction in the chosen medium, such as conventional books, microfilm, 'electronic books' with low-cost solar-powered Perspex flat screens, video, or software. Pilot projects have been exploring such potential uses of communications and information technologies for some years. Their successful exploitation depends on a sound infrastructure of technological and educational provision both in the original producing context, and internationally if the technology is to reap its full potential benefits. This implies a level of long-term investment that is not popular in the United Kingdom, which nevertheless has been the source of many of the component technologies at the research and development stage.

All these technologies are already being exploited within every major financial centre. The City of London has installed integrated advanced information technology at enormous expense, to handle tens of thousands of financial transactions each day. To outsiders it seems uncertain whether all this effort creates a better world. The pace of change is unconcerned with the human scale. Some critics saw the 'Big Bang' as a prelude to the wholesale destruction of the traditional role of the City of London. They warn that the full impact of harnessing new technology has yet to be felt, citing the automatic selling of shares computerised traders in New York, which wiped billions of dollars off the world economy without any human intervention in the form of professional stockbroking advice.

We can add another level to this developing picture of technological change. Advances in speech input technology mean that commercial systems are already available which, one the system has become familiar with the speaker's voice, can be used to give spoken commands to computer systems. Other systems incorporate speech chips which enable the system to give an audio output that approximates to normal human speech, even taking account of regional accents. Optical character recognition system currently available allow the computer system to scan printed text, or even, in certain cases, handwriting, automatically transforming the visual image to computer-intelligible form. Systems now in use with the blind allow the user to have Braille input to and output from computers. Specially designed input devices for the handicapped such as the *concept keyboard* (which allows the user to touch pictures or words on the keyboard to denote particular responses), and the Possum typewriter (which allows input from any controlled movement such as blowing or pointing), support different modes of use, as do many available variants on the now-traditional but objectively eccentric QWERTY keyboard. The original inventer of that keyboard took account of the regularities with which certain letters were used in combination in English texts, and arranged the keyboard to ensure that the internal mechanical moving parts did not become entangled. That is to say, he arranged the keys in such a way as to keep the pace of the typist behind the capacity of the machine to type. Such a rationale is no longer valid, not least because modern information technology has few mechanical moving parts, but old habits die hard. In teaching it is still the case in some institutions that keyboard skills for computer users remains no more than an extension of typing.

The integration is already deeper than the level of component technologies such as printers, and has broad implications. Many personal computers now offer integrated packages that allow the user to combine word processing, databases and financial spreadsheets. Few can be said to be used very intelligently, as the overlaps and communications between the modes of computer use are poorly understood — integration is still more the name of a problem than of a solution.

Integration at the level of technology and process logically depends on the existence of an overall frame of reference, a unified view of the interaction between man and machine within which different combinations and configurations can be accommodated. It has been the argument of this book that such integration must occur at the level of knowledge, and involves issues which by definition transcend the narrow confines of the histories of the component technologies to date. We dare to offer such an overall frame of reference from the human point of view, within which we hope to make sense of the recent flurry of developments from the machine point of view, the most influential of which have occurred in Japan with the Fifth Generation Computer Project.

## 2.5 The Japanese Fifth Generation Computer Project

### Managing the future

Niccolo Machiavelli would have enjoyed visiting ICOT, the New Generation Computing Institute in Tokyo, the base of the Japanese Fifth Generation Computer Project. From here the Human Frontier Programme is intended to spring. He would doubtless have been invited as a visiting scientist, part of the extensive Japanese programme to draw on the ideas and experience of other research projects and institutions around the world. In Japanese tradition, copying is a form of flattery, and normally precedes adaptation and improvement: a form of syncretism.

The Japanese regard logic programming, which has been chosen as the basis of their fifth generation computer systems, as a form of martial art, an intellectual discipline that can be extremely awkward and uncomfortable. It was learnt at the feet of masters such as Alan Robinson, David Warren, Alain Colmerauer, Ehud Shapiro, Robert Kowalski and Keith Clark. As with the learning of archery through set martial exercises, regular practice of logic programming can induce reliable and accurate habits of thinking. It was a somewhat unexpected discovery that, apart from being good for the soul, logic programming can produce exceedingly good and efficient computing systems in the real world. Kazuhiro Fuchi, Director of ICOT, wrote, not entirely seriously, 'soon even cats and spoons will program in PROLOG.'

American research in expert systems technologies in the 1960s at Stanford University and Massachusetts Institute of Technology, was given considerable attention by Japanese researchers such as Kazuhiro Fuchi and Keochi Furukawa, who in the 1970s visited Stanford Research Institute, where they first encountered PROLOG. However, American work until 1984 was based on the list processing language LISP. Systems were constructed that allowed users to address real problems of knowledge, such as medical diagnosis and mineral prospecting. The emphasis on the facilities provided by particular hardware and software systems, and the evolution of complex 'programming environments' with powerful graphics capabilities has often been at the expense of the rigour of the work done. Programmers can do impressive things on a screen without understanding what they are doing or appreciating the limits of the capabilities of their system. This has made LISP-based systems notoriously difficult to transfer to cheaper conventional computers and microcomputers, which would have been necessary if LISP-based computing was to spread in Europe where equipment budgets are much smaller than in the United States. American LISP systems have also proved hard to adapt to new highly parallel computer systems, now under development and running in prototype form at Imperial College, Manchester University, Massachusetts Institute of Technology and ICOT in Tokyo.

The approach to knowledge that has been taken by workers in the LISP tradition emphasises the *procedural* aspects — how things are done. The logic programming approach, explored by Europeans since 1970 and preferred by the Japanese from 1981, emphasises the descriptive or *declarative* aspect — how things are, what the problem is. It should of course be understood that in order to solve a problem we need to be able both to describe what the problem is, and then to proceed to how it is solved. In the Fifth Generation Computer Project, description is seen as taking precedence, particularly as the ways of solving problems may be changed by the advent of computers with more than one processor, working collaboratively. Another common ingredient in the new systems is 'object-oriented programming', where computing is seen as a matrix of communication between processes. Amid all these powerful concepts, deriving from the long tradition of artificial intelligence research, it has been important to keep a clear head, and not just be dazzled by the cleverness of the surface appearance of the technology.

The Japanese plan was to challenge the IBM and American traditional stranglehold over information Technology by leapfrogging, by playing down the importance of fourth generation systems now being installed, and focusing on the more distant horizon of fifth generation systems which were going to be, quite simply, something else. Fifth generation machines would not only embody distinct advances at the level of VLSI technology (very large scale integration of the electronic circuitry used on the new generation of microprocessors) and microprocessor design, but would offer the opportunity for a radical change in directions and paradigms at the levels of computational model, programming languages, system development environment, applications and user interface. This was all far too difficult to be taken on casually, and did not simply represent an incremental step from the previous approaches to computing. It required a major effort of theoretical and practical work. Practical results would not be immediate but would be critical, as the ideas, though fundamentally simple, appeared not only extremely complex but in many senses threatening to the security and status of conventional computer scientists. Even within Japan, the new project would involve friction and conflicts of interests in the short term, which it was hoped would prove to have been justified in the long term.

An enterprise which had such broad ambitions, yet such a tight technological focus, required a major collaborative effort and novel techniques of research management. Laboratories at ICOT are laid out in a way which corresponds to the directions and intended relationships of their technological components. ICOT is a physical model of the kind of parallel approach, intelligence and communication facility that are required of the finished systems of the 1990s. This is no accident. The approach of problem-solving by breaking down a large problem into component sub-problems has long been a key tenet of Japanese management. As long as research managers do not make the mistake of thinking that real human problem-solving is always like that, they are likely to continue to enjoy considerable success, based on meticulous research and early products delivered to schedule.

## Technical emphases

The emphases of the Japanese Fifth Generation Computer Project were announced at the conference in Tokyo in November 1981, and have remained remarkably consistent. Kazuhiro Fuchi, Director of the project, reported in a keynote speech at the Third International Conference on Logic Programming at Imperial College in July 1986 (see section 5.2 for details):

'The FGCS Project started four years ago, but its roots go back three years before that. More than a hundred representative researchers in Japan participated in the discussions during those three years. A clear consensus emerged that logic programming should be placed at the centre of the project.

We announced this idea at FGCS 81 in the fall of 1981. It was the most controversial issue at the conference, criticised as a reckless proposal without scientific justification.

Why did we persist in our commitment to logic programming? Because we were inspired by the insight that logic programming could become the newly unified principle in computer science!

At the conference, Kazuhiro Fuchi pointed out that logic programming covers computer architecture, new programming style, semantics of program languages and database machines. It was also pointed out that logic programming is playing an important role in linguistics and artificial intelligence.

Looking at the situation now, we can say that our conjecture has taken more concrete shape. It is not too much to say that the successes of the project so far are in large measure due to our adoption of logic programming. We would like to emphasise that the key feature of our project is not knowledge information science or non-von Neumann architecture, but logic programming.

The results we have achieved seem to be quite natural. Therefore, it may be more appropriate to say that what we have here is a case of 'discovery' rather than 'invention'.

Our approach may be compared to the process of solving a jigsaw puzzle. The process of putting each piece of a jigsaw puzzle in the right place may correspond to the process of discovering truth in our research. Also, the complete form of the jigsaw puzzle corresponds to the highly parallel computer for knowledge information processing realised in VLSI. Logic programming is the centrepiece of the jigsaw puzzle. Logic programming is the 'missing link' connecting knowledge information processing and highly parallel computer architecture. One we realised that, the puzzle started falling into place rapidly.

We often hear people say that our project ought to be more software-oriented rather than hardware-oriented. But our perspective is the whole jigsaw puzzle, of which software and hardware are certainly inseparable aspects, but neither takes priority as such. We proceed step by step, with our vision of the new computer guiding our approach to the solution of each problem as it arises.'

## 2.6 International response

The Japanese proposal of a new generation of computer systems, and the invitation to the international community to collaborate in a joint programme of research and development, sent a tidal wave through governments, industry and academia, and focused attention in an unprecedented manner on policy issues in computing and information technology at a national and international level.

## Europe

European research in advanced information technology has played a central role in establishing the foundations of the Japanese Fifth Generation Project. In particular the work on logic programming and PROLOG by Robert Kowalski, Alain Colmerauer, David Warren and Keith Clark has been critical, much of it emanating from collaborative work at Edinburgh University in the late 1960s and early 1970s, also involving Alan Robinson, Pat Hayes and Alan Bundy. Many leading British researchers in the field were obliged to move overseas when the Lighthill Report in 1973 to the Science Research Council led to the removal of financial support for much of the British research work. The British government were less able to take a long-term view of research and development than were the Japanese a few years later, looking at fundamentally the same research.

The Japanese example of establishing new national collaborative structures involving advanced research and computing manufacturers has been followed widely. The European Economic Community has founded the ESPRIT Programme (European Special Programme for Research in Information Technology), led by the major European information technology companies, and subsequently EUREKA, a program initiated by the French President Francois Mitterand to apply European advanced technology research to civil applications in the European market. Academics, companies and governments have become accustomed to international collaborative work and, despite early teething problems concerned with bureaucracy and problems arising from differences in language and institutions, a new status quo has been established. To qualify for support under ESPRIT a project must be sufficiently 'communautaire'; it must involve collaboration between academics and industry in at least two EEC member countries. In the case of EUREKA participation is extended to many nations who are not members of the EEC, but who have relevant skills and commercial strengths to contribute. Such collaboration is not purely technical in motivation: it provides a concrete example of the benefits of international organisations, in terms of economy of scale, unification of efforts, and constructing a European strength in advanced information technology to rival that of the United States and Japan.

In general terms the technical emphases of ESPRIT are fully consistent with those of the Japanese Fifth Generation Project, but the programme is more diverse, less focused, and less tightly structured. This is perhaps inevitable when so many countries, groups and interests are involved. As the ESPRIT programme has developed the technical hard core has become more unified and apparent, massaged into shape by successive work plans produced by the Task Force based in Brussels.

The Japanese example of close collaboration between companies has been followed in the establishment of the European Computer Industry Research Centre in Munich in 1983, involving ICL, Seimens, and CII-Honeywell-Bull of France. The technical focus at ECRC has again been on logic programming, parallel computer architectures and knowledge-based systems, and the European dimension to long-term research and development, apart from producing new products for the European market, is having a healthy effect on corporate attitudes to investment, research, development and training in the long term.

Individual European countries have also made considerable progress in establishing national programmes in advanced information technology, typically having to make up for years of relative neglect by comparison with the emphasis given in the United States and Japan. Since 1983 the relationship between national and European programmes has developed and become better understood, to the extent that each member country now sees its future policy in terms of both national efforts in fundamental research and community collaboration in development and market exploitation.

In each of the member countries and in the European collaborative programmes there is a heightened awareness of the importance of appropriate education and training in and with information technology. Work on common technologies in research is beginning to engender common approaches to education and training policy. This is evident in the recently published *Information Technology and Education: The Changing School* edited by Richard Ennals, Rhys Gwyn and Levcho Zdravchev, which draws considerably on the experience of the association for Teacher Education in Europe. The base of relevant educational experience in Europe can be seen in *New Horizons in Educational Computing* edited by Masoud Yazdani and other collections books which are cited in section 5.2.

## The United States

For the United States the Japanese Fifth Generation Project has posed a challenge akin to the Russian launch of the Sputnik in 1957 and the first manned space flight in 1961. It has led to a fundamental reassessment of assumptions of American technological superiority and of the dependence of advanced information technology on funding from military programmes. Extremely painful questions have also been asked about the national education system, which is generally agreed to be in a state of crisis.

The initial American response to the Japanese proposals and their technical emphases bordered on derision. Complacent Californians comforted themselves with the thought that the Japanese choice of logic programming would delay all useful developments, and that the real advances were to be expected from the continuation of high-technology expert systems development based on procedural programming, software engineering techniques and standard languages such as ADA and LISP.

In the years since 1981 there has been a major change in American computer science, and logic programming is now embraced, together with parallel processing, as providing the basis for future developments. The emphasis on small companies, the entrepreneurial approach, and the lack of a collaborative culture with long-term planning outside the military-industrial complex has generally meant that many of the advances are more in terms of individual tools and techniques than, in Fuchi's terms, the overall 'jigsaw puzzle'.

Large companies such as IBM now devote considerable resources to fifth generation developments, and have recruited leading European researchers to head their new laboratories. Many of the major companies, though not IBM, have pooled their resources in the new Microelectronics and Computing Technology Consortium (MCC) in Austin, Texas, such a novel move that it required the amendment of American anti-trust legislation. Again we can discern the familiar research emphasis on logic programming and parallel computer architectures.

American government activity is focused through military programmes, which have traditionally supported research and development costs for a large proportion of their high-technology industry. The Department of Defense Software Engineering Study, the Strategic Computing Initiative and the Strategic Defense Initiative are concerned to develop advanced information technology in the United States and strengthen the competitive positon of American firms, irrespective of the technical feasibility or otherwise of the overall projects.

An analysis of the technical requirements for the next generation of computer systems, whether for civil or military application, shows a common 'shopping list', many of whose items are currently best developed in Europe. Such a list, to be found for example in the Innovative Science and Technology Program of the Strategic Defense Initiative governing contracts on offer to European universities and colleges, would include:

high performance
parallel architecture
reliability
artificial intelligence
real-time performance of expert systems tasks
speech input
powerful environments for users and programmers
robustness

large databases
the capacity to reason about changing information
facilities for explanation
interaction through natural language

## Governments

The development of a new generation of computer technology is a task that is beyond the resources of individual companies, or of the private sector alone, particularly where there has not been a strong tradition of long-term in-house research and development. Advanced information technology is so pervasive in its applications and implications that its management and development assumes a significance for national and international policies of economic and social development, and of security. A philosophy of 'laissez faire' is not tenable, or the result will be that the richest countries and companies can simply purchase control of the future. This is a critical new ingredient for politics and political science. Information technology is in itself neutral, but when applied it will reflect and reinforce the political and economic philosophy of its user.

## Companies

Internationally companies are having to address the issue of advanced information technology, which affects all of them. A business financial strategy based on cost-cutting and the closing down of research and development is likely to prove fatal for the companies concerned, even in the relatively short term. The pace of technical change is such that none can stand aside and be justified in feeling immune from its effects. As a simple example, the development of intelligent communications technology means that at the press of a button financial dealings can be relocated from, for example, the City of London, to a more appropriate environment anywhere in the world. Companies are now, whether they like it or not, operating in what Marshall McLuhan called a global village, with ever tighter pressures of competition which necessitate changes in practices and approaches.

## Education and training

If there are answers to the kinds of problems we have been exploring, they will be reflected for particular countries and companies in the use that they make of education and training. We may not know, or even care about, the details of the new generation of computer systems which are under development. We know that the new generation of users is currently in our systems of education and training. Cuts in education and training threaten the economic and political future of a nation, and cuts have been made for many years which may unfortunately prove to be irreversible.

## 2.7 The British Alvey Programme

The first response of the British Government when invited to participate in the Japanese Fifth Generation Project was to commission a report from a committee chaired by John Alvey of British Telecom, looking at the problem addressed by the Japanese and considering appropriate approaches. The unanimous report of the Alvey Committee was submitted in September 1982, and led to the establishment of the Alvey Directorate in March 1983, following the majority of the recommendations of the report. The principal change was that government funding of advanced research by industry, recommended to be at a level of 90%, was reduced to 50%. A national programme of research and development involving government, universities and industry was recommended, with a budget of £350 million over five years, of which £200 million was to be from government.

### Outline of the programme

The Alvey Report was unanimous in its recommendation of collaborative research and development in the enabling technologies that would be required to underpin the new generation of computer systems: intelligent knowledge-based systems, man-machine interface, software engineering and very large scale integrated circuits (VLSI). Accordingly, separate directorates were established to deal with each of these areas, responsible for encouraging new collaborative projects, with the objective of progressive integration of the programme and the establishment of links across the arbitrary divisions between, for example, software engineering and intelligent knowledge-based systems.

In additon, large demonstrator projects were established which were intended to draw on the advanced technology for each area and develop it into a visible commercial application, each led by a major information technology company. Some communications infrastructure was provided and a modest degree of administrative support.

The Alvey Directorate was small, drawn largely from industry and government, but with an increasingly significant role being played by the universities, upon whose research advances the programme sought to build. The Directorate have worked as catalysts and 'marriage brokers', introducing prospective collaborators and assisting in the complex process of arriving at legal collaboration agreements and contracts covering the conduct and exploitation of the work.

It be became clear very early in the life of the Alvey programme that research and development was contingent on the development of a growing body of skilled researchers and practitioners. The Alvey

Report stated that in the critical use of intelligent knowledge-based systems (including logic programming and parallel computer architectures) there were only some four dozen experienced workers in the United Kingdom. Many leading researchers had been tempted overseas by greatly superior salaries and working environments in the decade following the Lighthill Report in 1973. Many Alvey projects have had to rely on a very slender basis of expertise, and have really had the character of training exercises. The Intelligent Knowledge Based Systems Directorate has conducted, in addition, an extensive awareness programme, providing an 'Expert Systems Starter Pack' software, videos, secondment schemes for industry and 'community clubs' which have brought together companies with similar applications needs to work on the development of demonstration systems, in areas such as insurance, process control and the transport industry.

In April 1986, 303 projects had been approved, of which 187 were full industrial collaborative projects, and 116 involved academic participants only. 109 firms, 53 universities, 11 polytechnics and 20 other research establishments were involved in projects, with an average of 3.9 partners per projects, consisting of 2 to 3 firms per project and 1 to 2 universities per project. Some degree of concentration occurred in leading firms and universities as centres of excellence developed.

The following tables, taken from the report of the director of the Alvey Programme, Brian Oakley, to the Alvey Conference in July 1986, give an impression of the relative balance of efforts in the programme.

**Figure 1** Industrial participation in the Alvey Programme, May 1986

| Firm | number of projects |
|---|---|
| GEC | 57 |
| ICL | 38 |
| British Telecom | 36 |
| Plessey | 35 |
| STC | 31 |
| Ferranti | 16 |
| Logica | 15 |
| Systems Designers | 12 |
| Software Sciences | 9 |
| Thorn-EMI | 7 |

**Figure 2** Academic partners in Alvey projects

| Area | number of academic partners in project | number of industrial projects |
|---|---|---|
| VLSI | 18 | 56 |
| software engineering | 7 | 31 |
| IKBS | 4 | 43 |
| MMI | 0 | 40 |
| Infrastructure and Communication | 0 | 2 |
| Large Demonstrators | 0 | 4 |

**Figure 3** University participation May 1986

| University | number of projects |
|---|---|
| Imperial College | 34 |
| Edinburgh | 31 |
| Cambridge | 24 |
| Sussex | 16 |
| Loughborough | 16 |
| Manchester | 15 |
| Strathclyde | 15 |
| Oxford | 13 |
| University College, London | 12 |
| Southampton | 10 |

Brian Oakley summed up the achievements of the Alvey Programme in a summary chart, on which the last item deserves our particular attention:

*Alvey Achievements*

Technical progress
UK firms brought together
Universities and firms working in close partnership
build up of UK research community
'Bring back' of outstanding experts from US
trained many postgraduate research workers in shortage areas
created common strategic objectives for the UK community
established preferred standards for languages, equipment, etc.
created awareness of new technologies
focused attention on manpower shortage

# 2.8 The fifth generation five years on

## Progress report

The Japanese Fifth Generation Project at a conference in Tokyo in 1981 published a schedule of objectives over a 10 year period, and naturally chose to achieve implementation of basic enabling technologies in the early years. The Japanese objectives were initially misrepresented internationally, as some commentators talked in terms of 'thinking machines' within five years.

To the extent that specific dates were set for the delivery of finished systems and components they have been met, and progress has been made in new software technology based on logic programming. Use has also been made of research results in functional and object-oriented programming, which have been progressively integrated into overall systems. A Personal Sequential Inference machine (PROLOG machine) and the Delta Database machine are now in use, and advances have been made, with international collaboration, in the development of parallel logic programming languages and applications building environments. A number of experiments have been conducted with non-von Neumann parallel architectures.

In the field of applications the greatest emphasis at ICOT has been on natural language processing, using both sequential and parallel systems. This is critical for the production of the intended automatic translation systems which are among the principal objectives of the project, as conventional systems have failed to give the necessary sophistication or speed of performance. Related academic groups have pursued research in expert systems applications in medicine and other fields.

Apart from progress at the ICOT collaborative centre, where researchers from the eight largest Japanese computer companies work together, considerable efforts have been devoted in the research laboratories of the individual companies to developing their own products. We are beginning to see the first hardware (the PSI Sequential Inference Machine) and software (the ESP language system) reaching the market.

In Europe the ESPRIT Programme has committed the bulk of its resources for the initial five year period, and plans are being made for a further phase. Lessons have been learnt from the evaluation of the first phase, which was a novel experience in international collaboration and management techniques, apart from the technical content of the projects. The general response has been favourable, though a relatively modest budget has been spread very thinly over a large programme which lacked the form of the Japanese initiative.

In the United States the computing aspects of the Strategic Defense Initiative show signs of being influenced by the successes of Japanese and European computer science researchers, who have been offered contracts with the American Department of Defense. This provides evidence of the external respect for the work, which was intially treated with little respect in the United States where continuing technological dominance was assumed. Major computer companies such as IBM, and the collaborators in the MCC Consortium at Austin, Texas, have sought to follow similar directions in research and development, but appear to have been hampered by less imaginative research management and preoccupations with short-term financial considerations. IBM have launched new expert systems products for their range of mainframes based on PROLOG and LISP in 1986, and no major manufacturer can afford to be left behind. Issues of ownership of research and control over development and applications are far from resolved. The same underlying technology of intelligent thinking is required in order for a whole range of applications to succeed. There have been controversies over industrial espionage, 'dumping' of Japanese microprocessors, and United States Pentagon control of the products of militarily-funded research programmes. Advanced information technology is now a key component of modern politics and economics.

In Britain the Alvey Programme has succeeded in mobilising a large number of collaborative projects, breaking down barriers between organisations and research fields in a manner that cannot be reversed. The initial budget is fully committed to a set of projects in fulfilment of the original objectives, and the first by-products of Alvey projects are reaching the market. The effects to date are most marked in the management of research. Fundamental research in advanced information technology has been reduced as industrial collaboration has been a prerequisite for government support of a research and development project. At the time of writing there is a funding vacuum which could have calamitous consequences; research teams built up over many years can be scattered in as many days in the absence of continuity of funding. The availability of military funds for the field only serves to distort the balance of projects and to cut across the tradition of peer review of proposals on scientific merit. Collaborating companies have learnt some habits of long-term strategic thinking, but have continued to be reluctant, by comparison with their overseas competitors, to commit their own financial resources to research and development in the long term. Universities are left in a position of distressing uncertainty, unable to ensure the finance or infrastructure support for approved research projects. That advance planning which is the hallmark of Japanese programmes is conspicuous by its absence in Britain.

## Technology transfer

Technology transfer is the crux of the problem of advanced information technology, and the central issue addressed by this book. The 1982 Alvey report in the United Kingdom identified a grave shortage of experts in the enabling technologies, and it takes time to make a significant increase in their numbers, hampered considerably by continuing cuts in government support for education in general, and in universities and polytechnics in particular. Where support has been given for advanced information technology it has often been at the expense of related subjects.

Concern has been expressed at ICOT in Japan, where researchers are seconded for finite periods of years from companies which are increasingly reluctant to release them. These researchers regret the lack of close connections with the university system, which they have identified as an attractive feature of the British Alvey Programme. This means that the insights gained in advanced research at ICOT do not find an immediate place in the training of undergraduate and postgraduate students, and that the gap between industry and university life remains, contrary to popular myth here concerning 'practical' Japan. In conspicuous contrast to current attitudes in the United Kingdom, there is an increasing emphasis on the importance of fundamental scientific research in Japanese universities, from which later generations of technology will spring.

In Britain, partly due to the lack of past emphasis on government support for education or on industrial support for training, the Alvey Programme is also expected to 'pull through' much of industry and the education system into the intelligent use of advanced technology. There is the assumption that the same leading researchers can teach students, offer consultancy to industry, and participate in conferences and training programmes. It requires ingenious management for all of this to be possible without sacrificing the research itself, and killing the goose that is supposed to lay the golden eggs. Techniques of secondment, distance learning materials and regional information technology development units are being tried, but with a budget that suggests caution from government rather than wholehearted commitment.

Paradoxically, the key technology that needs to be transferred concerns the use of our brains in new ways of thinking, with or without the assistance of computers. Now is not the time to immerse students in conventional courses in programming and the intricacies of the working of microcomputers. Such courses are now of largely historical interest, and focus on the ways of thinking of computer science 20 years ago. The new challenge is to find ways, using existing equipment and finite budgets, of introducing appropriate ways of thinking involving advanced information technology into current practice in education and training.

The authors are co-founders of the Information Technology Development Unit at Kingston College of Further Education, whose aim is to assist in bridging the gap between advanced research in university groups such as that based at Imperial College, and the real world of industry, commerce and education. To this end the Unit collaborates in research projects using the technology in practical applications, and develops courses in knowledge-based computing for use in education and training. Close links are maintained with associated projects internationally, some of which are described in the following chapters.

Specifically, the Information Technology Development Unit is participating in an Alvey project which includes Logica, Imperial College, Exeter University and the Engineering Industry Training Board. The aim is to produce a computer-based training package employing interactive video, an intelligent tutoring system for CNC milling machine users.

The following chapters of this book provide introductions to a variety of practical applications of the technology, but cannot hope in the space available to provide detailed tutorial support.

# 3 Knowledge Exploration in Education and Training

## 3.1 Knowledge and problem-solving

### Ways of thinking

In focusing on ways of thinking with computers in the classroom, we are making certain assumptions, whether our emphasis is on procedural thinking using languages such as LOGO, declarative thinking using logic and PROLOG, or object-oriented thinking using SMALLTALK or its derivatives. Yet there is considerable common ground which unites the advocates of what have sometimes been seen as competing approaches.

Conventional education has emphasised the learning of factual information, and classroom lessons have been dominated by the prospect of the formal written examination, in which the candidates have been required to reproduce portions of learned material. Skills of memory and selection are therefore rewarded, understanding not infrequently being an optional extra. In order to make intelligent use of computers in the classroom we as teachers have to be prepared to abstract from the particular detail of a given subject area, uncovering the general ways of thinking that are involved in achieving expert levels of understanding. This in turn assumes a high level of understanding by teachers of their subjects.

Conventional classroom teaching is heavily directed by the teacher, who may well have 'given the same lesson' year after year. In order to establish what has been taught and learnt in a particular lesson it is merely necessary to have sight of the teacher's notes, which during the course of the lesson should have been reproduced in the pages of the student's book. When we are focusing on ways of thinking we are more concerned with the process than with the product. Each student will have thought in slightly different ways and more individual differences will be noted in written work and classroom behaviour. The classroom activities should be more student-centred, concerned with the provision of learning experiences for the individual learner.

It will be apparent that different uses of the computer in the classroom can be taken as signifying different approaches to education which long predate the advent of computer technology. In terms of educational psychology we can contrast the mechanistic behaviourism of B.F. Skinner as well as the arbitrary and fraudulent classification of children's capabilities of Cyril Burt with the cognitive development approach of Jean Piaget and Jerome Bruner.

We are concerned in this book with the practical implications of the theoretical insights of Piaget in his voluminous writings over fifty years, which has been appallingly distorted when introduced to teacher training students. It is often presented in terms of content itself, rather than as an account of ways of thinking rooted firmly in Piaget's personal experience and observations, and made coherent in terms of the structure of the problem domain and the relationship of the learner to what is being learned.

Jerome Bruner, in practical curricula such as *Man: A Course of Study* (see section 5.2 for details) has taken the formal models of cognitive development and understanding and added an appreciation of the implications of different modes of representing knowledge at different levels, and has suggested a variety of 'experience centred' curricula. He has written of the importance of iconic and enactive modes of representation (pictures and simulations), in addition to conventional text-based approaches.

Such approaches run the risk of seeming unduly abstract. A particular attraction of the computer is that it can give concrete form to abstractions. A program is the realisation of an idea, which becomes available for scrunity, manipulation and testing. It should be evident that this offers considerable attractions for the Piagetian teacher, who is concerned to enable his students to advance from concrete operational thinking to formal operational thinking. With intelligent computer the formal can be made concrete.

There are some instructive comparisons to be made between conventional CAL (computer-assisted learning), CBL (computer-based learning), CAI (computer-assisted instruction), and an approach to the educational use of computers based on ways of thinking, or AI (artificial intelligence). The former show a greater concern for particular details and presentation, and often aspire to being 'teacher proof' as one of their purposes is to replace the teacher as the source of wisdom in the curriculum. Computer sessions are typically very directive, with the student being led along a linear path of activities, encountering multiple-choice questions to which there is a correct answer. This is no place for the intellient sceptic, for the student who wants to ask why.

Concepts of 'ways of thinking' can seem vague and abstract if introduced with conventional media into the conventional classroom. The computer enables us to give a visible form to such ideas, and outline structures with which we can support the complex forms of our problems. What is more, computer programs can be used interactively, demonstrating what they mean, laying themselves and their users open to scrutiny and to question.

## Procedural thinking with LOGO

Although the primary aim of this book is to illustrate new views of computing, it is also important to focus on the effective aspects of existing computer systems when introducing these ideas to students. LOGO was in fact developed by Seymour Papert, from the artificial intelligence language LISP, as a means of creating an environment in which children could think and express their thoughts about mathematics. It is a procedural computer language in that a program in LOGO is a sequence of instructions telling the computer what to do. Its importance lies in the clarity for the user; even young children are able to have a clear understanding of what is actually going on.

The core idea of LOGO is the control of a simple robot or a turtle-shaped symbol by the program, which directs the movements of the turtle on the screen. One of the principal advantages of LOGO is the small number of commands the user needs to learn and the way in which they encourage structured thinking from the very beginning. The initial instructions simply tell the computer to move the turtle forwards or backwards, to the right or left, to pick up a pen and put it down again. These are the simplest procedures that a computer is able to perform. But soon the user can learn to combine these procedures in order to perform higher-level actions, such as the drawing of a square, a house or a whole street of houses. By this procedure a very small set of concepts may be built up into a set of higher concepts endowed with greater power. If we had learnt to do the same thing with every computer language, then current programs would be much better in that their users should have the in-built facility to reduce the high-level problems they are trying to solve to a smaller and more manageable size.

LOGO has the advantages of simplicity, easy access and a structured approach. It is capable of tackling complex ideas from a relatively humble base. That its inventor did not see it as a computer language at first, but as an aid to mathematical thinking, probably explains the ease with which people can use LOGO. Certainly Papert's intention was to encourage children to think about mathematical problems without knowing they were doing so. In his book *Mindstorms,* he makes his educational concern quite explicit and in consequence his discussion of computing provides a very useful starting point for teachers and trainers concerned with introducing new computer concepts.

One point needs to be stressed, however. Turtle graphics should never be equated with LOGO, since this language contains other elements which we have not treated. It possesses, for instance, the ability to manipulate both words and sentences. Another point of caution is the number of educational packages that describe themselves as being similar to LOGO; few of them equal what we should reasonably expect from a full implementation of the language.

## Object oriented thinking with SMALLTALK

Object-oriented programming represents another way of preparing ourselves for using computers in parallel, the norm of fifth generation computer systems.

The idea of object-oriented programming may be grasped by the analogy of a play on the stage. Imagine that a group of actors has gathered for a performance, each one having a character, a role to perform and lines to speak. Say the chosen play was *Snow White and the Seven Dwarfs.* Whilst Snow White is quite different from the Seven Dwarfs, there are still sufficient differences between the characters of the dwarfs themselves to provide a variety of roles and actions for all the actors. So at various points in the plot each actor will have an individual part to play, according to certain cues given by one or more of the other performers. It is essential that everyone on stage, or standing in the wings, or waiting backstage, is fully aware of what their own role is in the entire process — the dramatic performance.

Now object-oriented programming operates in the same co-ordinated manner, not unlike the words and moves detailed in the script belonging to the play's producer. It commences with an attempt to describe the objects, or entities, involved in the computer problem-solving process. Then it describes the environment or world in which these objects exist and explains the roles they have to perform in order to solve a particular problem. Like actors, the various objects will not necessarily be involved all the time but only be activated to do something at an appropriate moment. The signal to come alive and perform a role in the problem-solving process will be when the other objects within the environment have indicated that they should play a part.

We need to note that all these objects possess distinct characteristics. Just as players on stage, they bring their characters to bear on what is in progress and provide cues which set off further action.

An example of object-oriented programming might be a simulation car wash (figure 1). To start writing the program it is necessary to establish the objects involved in the process. At its simplest they could be cars, a washperson and a drying person. These may be all the objects we shall require within the particular object-oriented problem-solving environment we are simulating. Cars of course can do a number of things here. They can be waiting to be washed; they can be being washed; finished or even driven away. Similarly, the washperson can be waiting to wash a car; washing a car or waiting to wash another car. The same is true for the drying person. It is obvious that there has to be some communication in the simulation for the car wash system to work. What is necessary is that when there is a car waiting and the washperson is waiting, the washperson should start washing the car. When the washing is completed, the washperson can tell the car that it is ready to be dried.

**Figure 1**

SMALLTALK is one of a number of object-oriented languages that have been developed during the past two decades. The basic idea is even older and elements of object-oriented programming are found in many languages. One of the things that has made SMALLTALK so interesting as a computer language is that it was designed with a programming environment which is at least as important as the programming language itself. This permits the user an unusual degree of control over the environment in which the actual programming takes place.

Perhaps within fifth generation computer systems the potentialities of object-oriented programming will soon be realised, since there it should be possible for every 'object' to have its own process — in effect, its own computer.

## Declarative thinking with logic and PROLOG

The importance of PROLOG has been as a harbinger of declarative programming. It was the first language to allow the realisation of a new knowledge-based style of computer programs, but we should not therefore associate declarative programming exclusively with PROLOG.

Declarative programming is in essence descriptive programming. Instead of the procedural approach of 'this is how you are going to solve the problem', the declarative method poses the problem itself: 'Here is a problem I would like to be solved' — and leaves the solving of it to the system.

Although people have long wished to have computers tackle descriptive problems, this was impossible without a computer language which was readily accessible. Once it was appreciated that formal logic could provide the beginnings of such a language, and that logic in itself was relatively easy to computerise, the declarative approach became possible.

The first use of PROLOG in English schools was in Wimbledon, where Richard Ennals concentrated on the logical aspects of the language. Because PROLOG has to run on existing computer systems, it encompasses a large number of facilities which are not 'logical', but at least they allow the user to treat it as something very close to ordinary speech. Yet its greatest benefit perhaps is the ability to view independently the rules or statements put into a program. Just as object-oriented programming enjoys a separation of the objects involved in the problem-solving process, so declarative programming in PROLOG permits an independence of thought that clearly has implications for parallel processing. For in a declarative programming system, each rule or each statement is entirely separate and independent of any other, though they may relate with each other in describing a problem as part of the problem-solving process.

Many different versions of PROLOG now exist. Not all of them are easy to use, although in the research stages of development of both PROLOG and SMALLTALK great emphasis was placed on user-friendliness. The current version of PROLOG, micro-PROLOG, has still a considerable way to go to match the simplicity of SMALLTALK. This shortcoming derives from the fact that we have still not completely envisaged the final shape of the computer language, unlike SMALLTALK and LOGO. What is really needed is a simpler PROLOG program.

An advantage of PROLOG over LOGO, nonetheless, is an ability to handle problems in the humanities. LOGO is excellent in its original mathematical domain, and has been successfully extended to some areas of physics, but it is difficult to anticipate a more general use. Here PROLOG scores as an interdisciplinary computer language, despite weakness in graphics. Projects are currently known to be underway in history, geography and economics. An effort is being made too to find suitable ways of teaching staff and students some of the concepts of PROLOG programming without having to explain its full syntax.

Typically a PROLOG program, or a logical descriptive program, requries a high-level description of the problem to be solved. Rules often comprise such a description, as in an expert system. There may be several rules describing the same problem as well as other rules describing the conditions affecting the definition of these rules. Eventually a database of descriptive facts may be added. Development work on expert systems has shown that within the right environment one can actually begin to run problem-solving programs without having defined everything involved.

## 3.2 A Reflective approach to learning and teaching

### Logic as a computer language

Logic is increasingly recognised as playing a central role in advanced information technology, underlying many aspects of computing other than programming. In educational theory one of the justifications for the teaching of logic was that it could offer a valid concrete extension of work done traditionally in the teaching of latin and geometry, where the emphasis had been on formal patterns of thinking rather than content. A classical education was seen as benefiting from such a formal component, together with attention to issues of language and cultural understanding. Latin could provide a well-formed foundation for a diversity of later studies. In recent years it has appeared an anachronism, and, together with the diminution of formal aspects of English language teaching, its departure from the curriculum has left something of a vacuum at the core.

Logic, as emphasised in the continental philosophical tradition, cannot be properly divorced from its domain of application. It has been seen by writers such as Michel Foucault as underlying the different domains of discourse, providing the means of articulating concepts. In the Anglo-Saxon tradition of the philospphy of education Paul Hirst and Richard Peters have emphasised the logic of different 'forms of knowledge'. There has been a good deal of pedagogical rhetoric, but theories have all too rarely been put to the test. Logical descriptions which can be run as programs offer this possibility.

In the PROLOG study conducted in Wimbledon middle schools, the original pilot class having developed a simple facility in the use of the formalism of predicate logic, showed little hesitation in applying their new tools to subjects of their own choice. Given that children use the same brains in the full range of subject lessons, it is natural that they should seek to apply ideas and techniques across conventional subject barriers, in the manner encouraged by Piaget and Papert. It can be argued that experimental observations of the phenomenon of 'decallage', whereby the level of formal operational thinking is supposedly achieved at different ages in different school subjects, tells us more about the way in which the different subjects are presented, represented and taught, than about the potential for children's learning.

Schools and teacher training institutions have shown a marked reluctance to transcend conventional subject boundaries, except in pilot courses which have typically been integrated only in name. Even in further education, where examination bodies have insisted upon an integrated approach, subject specialism has died hard. Intelligent computing will necessarily offer an opportunity to change that; we should not underestimate the implications for the future of education and its institutional structures.

## The humanities

The present authors have backgrounds in the humanities, and share a concern to explain a complex world to others, teaching their students how to understand and participate in the structures and processes that they study. Essential to this is reflection on how we think, how we reason, how we draw conclusions from evidence, how we communicate our conclusions, how we cope with conflicting information and incomplete knowledge. The real nature of the subject needs to be experienced in the learning environment. The methodological approaches of the new history, the new archaeology, the new literary criticism and so on need to be combined with a pupil-centred and experience-centred and approach to teaching and learning.

Specialists in history and the humanities have often been reluctant converts to the use of information technology. Previous generations of computer systems, which obliged the user to express problems in primarily numerical terms, seemed alien to the humanisitic tradition.

There have been exceptions to this rule, notably from those who have pursued a structuralist approach to history, the humanities and the social sciences. Such people have long used models to explain complex phenomena, not in the hope of providing a complete mechanistic working system to replicate the real world situation, but with the intention of aiding understanding.

When we talk of the feudal system, or of revolutions, we are making use of organising or colligatory concepts, simplifying generalisations which serve to point up interesting differences and exceptions. When we explain how things are, or how they have been, we are not committed to deterministic predictions as to how things will be. We are accustomed to describing the same set of events in different terms in order to illuminate connections and associations, or to establish common ground with our interlocutors. No historian ever believes that he knows the full picture; it is not possible to talk of complete knowledge of the past, rather we have to make use of our past experience and knowledge in general, together with the incomplete information we have about particular problems.

It should become apparent that quantitative history in the sense pioneered by Peter Laslett in Cambridge drawing on the work of Louis Henry and the Annales School in France depends upon a preparedness to identify important parameters through which we can come to an understanding of complex events. For quantitative history to be effective some of the chosen explanatory parameters need to coincide with consistently available sources of information. A study of social history that depends on a study of the institution of the family is fortified by the availability of census data, wills, legal depositions or records of inventories. As Laslett has pointed out, the very activity of generalisation that is central to the work of the historian or social scientist, organising facts under rules, assumes the use of existential and universal quantification, though not necessarily the use of numbers.

So in the classroom at Bishop Wand School in Sunbury, a course in local history draws heavily on primary sources concerning the village over the centuries. Young student historians need to make sense of a mass of information, building up and testing their hypotheses. Given the representation of organised bodies of records as databases the computer can assist this process, performing the chore of searching and correlating according to the description of the student.

A critical element of our humanity which has been poorly understood and undernourished in the United Kingdom is language. It may be to our ultimate disadvantage that English is widely spoken internationally, and the principal language of literature and the media, so that there is comparatively little pressure to learn foreign languages out of necessity. It can be argued that the power of one's own natural language cannot be appreciated without experience of learning other languages and of being obliged to communicate in them, to use the language correctly in real-life situations. Among the languages now on offer are artificial computer languages, which, given the appropriate choice of language and approach, offer analogous experiences. In this sense an appropriate language must offer facilities to describe information and procedures, and to build models of knowledge structures and grammars, rather than simply addressing the internal workings of a machine.

It is no coincidence that PROLOG arose from the work of Colmerauer in developing natural language translation systems in Canada. Indeed, in the same province of Quebec in which the political problems of bilingualism motivated the research into automatic translation, the University of Montreal is now concerned to teach French with the use of computer expert systems.

## Creative writing and literature

It is a mistake to contrast structured thinking with creativity, for each is impossible without the other. Creativity is required for the process of discovery, which is then rationalised retrospectively for the process of justification in scientific structures. Without a sound structure of knowledge we cannot appreciate and make sense of individual actions or utterances, themselves the product of the structures of the human agent.

When we study literary style, the devices employed by the poet or novelist to achieve his desired effect, we are in effect studying the work of knowledge engineers, the procedures and techniques that have been used to exploit the parallelism of a given knowledge domain. Devices such as alliteration, assonance, simile, metaphor, analogy, and so on are all concerned with forms of parallelism reduced to the sequential form of the written word on the page (sequential when read aloud, analysable as a whole on paper in a highly parallel form).

It is somewhat fanciful to suggest that literature is the trace left by past knowledge engineers making their path through adventures of ideas, forests of doubt, plains of despondency, mountain ranges of excitement. Such a view assumes that *Homo sapiens* is the engineer, understanding what he is doing rather than just, like our predecessor *homo habilis,* doing things with tools. But literature, and literary activity, does presuppose a society, with shared language, assumptions, objectives and communications.

It is worth recalling that Ludwig Wittgenstein, the enormously influential linguistic philospher, was a trained engineer. He explored the concept of collections of linguistic tools in a toolbox, to be taken out and used on particular occasions. His unifying concept was the game, played within a form of life, by people whose innermost feelings are unique, inexpressible and inexplicable.

## Mathematics

It is insufficiently understood how mathematics operates as a symbolic language system, transcending cultural differences. It underlies all areas of our experience, and requires us to take on the concept of working with formalism and abstracting from reality.

There are dangers in the apparent ease with which mathematical reasoning is susceptible to computerisation. Lessons can be learnt from the widespread use of BASIC in the teaching of mathematics, where there is a fundamental inconsistency of approaches. For example, the use of variables in BASIC is dominated by the working of the underlying computer to the detriment of pure concepts of mathematics. It is quite easy to confuse in this case mathematical equations with machine instructions.

## Science

Science must be rescued from its near-fatal entanglement with technology. Science is about asking questions, while technology is about providing answers. Answers without preceding motivating questions are peculiarly dangerous. The fundamental purpose of research, with a computer or without one, must be the establishment of crucial questions.

Scientific computer programs raise further problems in the hands of a third party user. They embody the writer's answer to a particular question which faced him on a previous occasion, which is rarely spelt out with the documentation which accompanies the program. They assume a particular approach to a problem domain, a way of classifying the phenomena in that domain, and a model of scientific reasoning, none of which are typically open to scrutiny when the program is used. Typically the program will be used as a 'black box', into which a form of input is given, with the result of a form of output, without an explanation of what has gone on in between.

There is a tendency to accept answers from a computer that would not be accepted from a human scientist, to oblige the user to suspend his scientific critical faculties in order to use the technology. In order to use a computer system you have to assume the correctness of the model, or belief system, which it embodies, and it is all too easy to forget that it is simply a model and cannot ever be a full reflection of the real world. In a simulation system adjustments are made to individual parameters, and it is assumed that all other elements remain constant. The real world is not like that, but computers, by their physical existence in the real world, give the illusion of operating on the real world. This can never be true in the full sense.

We have to recall approaches to science that were current before the advent of advanced information technology and the intellectual division of labour. Scientists must be freed to ask questions whose answers require a search beyond the conventional boundaries of the established specialist disciplines, or science is by definition condemned to stultifying inertia.

One exciting approach is that of naive physics, pioneered by Pat Hayes and pursued further by Jon Ogborn of London University Institute of Education. Under this approach we try to model 'common sense' concepts of science, trying to come to terms with the assumptions which are normally left implicit. How are we to describe the phenomena of motion, for example, if we set aside the underlying assumptions of Newton's laws? Can we begin to understand the astronomical concepts of the past, with their elaborate structures of spheres and cyclical motions? Here we are addressing fundamental questions of the history and philosophy of science, not at the level of postgraduate studies, but in a form that could be grasped by secondary school pupils. The computer can give concrete form to abstract concepts and display the dynamic implications of a theory which is usually expounded in static sequential text.

Complex ideas such as wave-particle duality are open to wider exposition through the medium of the computer, which allows us to choose our knowledge representation formalism for a particular problem. In this case the paradigm of object-oriented programming offers enormous potential for the teaching of complex molecular and biological structures, which can be seen at a number of different levels and explained accordingly.

### Engineering and technology

In the United Kingdom we maintain the myth of separating technical and vocational education from mainstream academic education. This is a colossal cultural error, which carries within it the seeds of the destruction of Western technological culture. It derives from the traditional class structure of society which when coupled with the industrial division of labour, has separated the skills of asking intelligent questions from the skills of taking practical measures to solve problems. Unless we can reunify the practice of education and training we are enforcing the divisions in society which, on Marx's analysis, will lead to its downfall in revolution.

With the computer the long overdue reunification can be radically assisted. From the same workstation the user can explore ideas and model real materials into physical products. Divisions of thinking and labour can be overcome, unless we choose to resist that obvious development.

## 3.3 Classroom toolkits for thinking

It has become possible in the 1980s to take a new higher-level look at the use of computers in education, liberating the user from concerns about the particular programming language that he or she is using, and providing an environment to support intelligent thinking in the classroom. A number of factors have combined to produce this situation. Technological advances have reduced the cost of computer memory, opening up new possibilities for the use of artificial intelligence languages and environments on personal computers and workstations. Work in the different languages LOGO, PROLOG and SMALLTALK has led to an increased awareness of the powerful ideas that can be made available for use by teachers and students. The research community has matured to the extent that it is now agreed that no one language or programming paradigm offers the answer to all problems, and a new generation of hybrid systems has appeared, purporting to offer the benefits of two or more of the strands of procedural, declarative and object-oriented programming. Most interesting of all, there is an increasing concern for the structures of knowledge. It has become clear from the experience of large-scale advanced projects that unless there is a firm foundation of understanding of problems of knowledge and knowledge structures, knowledge-based systems will be baseless.

In the world of commercial computing there has been a growing emphasis on software tools in the software engineering community, and on expert system shells in the knowledge-based systems community. In each case the vendor does not claim to have produced a full solution to your problem, but to have developed a useful tool which can be an aid to solving a new problem of an appropriate form. Typically such systems will offer a particular mode of interaction with the system, a motivating metaphor, an appropriate knowledge representation, support for a particular pattern of reasoning, and interfaces to other software systems. They should support the exploration of a new problem area, helping to cope with complexity and the 'housekeeping'. In the commercial world several such tools may be offered as part of an integrated system; in education we are more likely to encounter the tools in the toolkit singly, though we may well be able to move our problem around between tools, clarifying different facets of the structure.

## Adventure games

The spread of microcomputers has also engendered the spread of software for use in home entertainment, with a particular area of interest being adventure games. In conventional terms this involves the user in playing a game previously devised by someone else. In the classroom the tables can be turned and students can build adventure games of their own. This may be an extremely powerful and motivating way of exploring a subject, as long as the effort in developing the adventure is focused on the problem domain and not on the manipulation of machine code to produce superficially attractive visual effects.

The concept is not entirely new. Lewis Carroll's *Alice in Wonderland* and *Through the looking-glass* could very naturally be presented in the form of an adventure game, introducing numerous powerful ideas and concepts through the narrative and, crucially, humour. Another insight into adventure games can be obtained by considering the popular nineteenth century pastime of writing parodies and imitations of famous poems. The key characteristics were identified and satirised, and the familiar form was used to act as a vehicle for new ideas. George Orwell gave a similar analysis of boys' adventure comics in his essay on politics and literature 'Boys' Weeklies'. A conventional unchanging world view was presented which could help to mould the ideas of the young and impressionable.

One great appeal of the approach to adventure games taken, for example, by Jonathan Briggs with his PLAN program, is that it is to a large extent open-ended. The user - teacher or student - can develop his or her description of a situation further, and can check the system and its components at any time. Children and students are sufficiently familiar with the genre to be able to develop games of their own, organising their ideas and experience into one of a number of classical structures. The process is curiously reminiscent of the work of structural anthropologists, explaining the actions and practices of ancient civilisations in terms of myths and legends, with which aspects of the culture of the past can be brought to life. The cave paintings of early man, with their use of animal imagery to help in coming to terms with the world, are perhaps the true precursors of Dungeons and Dragons, Space Invaders.

## Simulations

Simulations can be powerful tools at a number of levels, offering models of aspects of the world as an aid to better understanding of the real world. There is a sense in which all lessons, all teaching, involve simulation, as we are always working on the basis of incomplete informaion, a known body of knowledge from which we hope to extrapolate to particular instances. Simulations can make these important logical points explicit.

Several things can be simulated: the nature of a problem area, the mode of interaction within a problem area, and the expert procedures of the specialist. Particularly in subject areas of great complexity or danger, or where considerable expense is involved in participating in the real activity, simulation methods can be very attractive.

Simulations are intended to further understanding, but can all too easily inhibit it instead. All too often simulation systems simply offer a numerical answer, and are unable to offer either an explanation or the facility to revise the program.

The underlying principles of simulation are simple enough, and given the appropriate tools, students can build their own simulation systems from a very early age, identifying decision points and outcomes of different courses of action. Standing back from a problem, seeing it in terms of its form, is a very high-level activity, yet the work of Jon Nichol and Jonathan Briggs has shown it to be within the grasp of even young children, given the right encouragement and the catalytic involvement of the computer system.

John von Neumann worked in the 1930s modelling human economic behaviour in terms of games and decision theory, drawing complex tree structures to reflect the patterns of choice and probability in, for example, two-person zero-sum games. The richness of the concepts involved far outstripped the capacity of conventional media to do justice to their power, and the printed page shows simply static diagrams. Other researchers also endeavoured to identify some of the underlying models and mechanisms behind complex economic phenomena. Their descriptions need to be run as computer programs for their true dynamic power to be unleashed. In the absence of computer programs based on predicate logic, such theorists turned to the inadequate substitute of probability theory and statistics, embedding their insights in a miasma of tests and approximations. We must return to the original writings and retrieve what they had to say, stripping off the statistical wrappings unless they too can be given a firm descriptive and logical basis.

We should be able to build computer simulations of complex theories in the fields of astronomy and medicine. In the process it is likely that we shall uncover common abstract conceptual structures. In the social sciences we should be given new insights into the ideas of major theorists.

## Databases

Databases are collections of organised information, assembled in a standard form. When addressing higher-level issues, and when organised in the form of facts and rules that make them open to interrogation and manipulation, they can be referred to as knowledge bases. As such they may, in printed form, be indistinguishable from the organised bodies of knowledge which we find in official records, the City pages of the *Financial Times*, *Wisden's Cricket Almanack* or *Who's Who*. In each case they will clearly be partial in their reflection of reality, governed by the questions that were asked of the real world when they were assembled, and the sources which were available to them. They need to be subjected to the closest scrutiny. We have to know who assembled the database, in what context, and for what purpose, and we must not fall into the trap of treating the information that we obtain from a computer database system with any more or less respect than that accorded to information derived from other sources.

Interaction with databases is a matter of the utmost concern for education and training. It cannot be assumed that everyone needs to develop a fluency in computer programming, but there are few who will not find themselves increasingly in contact with databases and information systems. Information about all of us is stored in numerous official and commercial databases, to which we may need to have access.

## Spreadsheets

Spreadsheets represent an increasingly popular tool in the business world. Figures and formulae are entered within a framework that models a commercial balance sheet or ledger. Totals can be automatically produced and rows or columns can be computed from others, using a simple mathematical relationship. For example, if we are computing the cost of manufacture of a component, and the raw material cost is increased, we can see that increase reflected across the board. Similarly, if a 30 % wage increase is agreed for all workers, such an increase can be applied automatically to all wage bills.

### Expert system shells

In Section 2.2 we outlined the concept of expert systems. Although expert systems have been around for a number of years, there are few that are commercially available. Those that are, tend to solve problems which have only a limited use within education. However, one recent development may be more appropriate to schools and colleges. Instead of purchasing a completed expert system, which is dedicated to solving a specific problem, it is now possible to acquire *expert system shells.* These are frameworks into which expertise about a particular topic or subject can be entered. Most will operate on the 16-bit computer systems now being marketed.

Despite the obvious attraction of expert system shells, they are still very much in their infancy. Almost all have been developed for non-educational purposes and a considerable investment of time will be required in order to explore their intricacies and their potentialities. We should beware of the slick commercial presentation that overlooks these difficulties.

At Kingston College of Further Education the IT Development Unit has a number of projects addressing these issues. As well as investigating the value of commercial systems such as APES, XI, ES/P ADVISOR and EXPERT-EASE, the Unit is developing its own range of simpler shells adapted to the needs of teaching. The examples given in Section 3.4 make use of these tools.

## 3.4 Expert systems in the classroom

At the time of writing there is little experience of the use of fully blown expert systems in the classroom. There are few complete expert systems in regular commercial use, though many are in preparation, and numerous prototypes are providing valuable lessons. It is not the argument of this book that large-scale expert systems are to play a major role in education and training, but rather that expert systems ideas can be highly influential. As we have argued, many of the ideas behind expert systems are not entirely new, but have long academic pedigrees, and their application goes beyond the use of computers. Artificial intelligence and expert systems involve a process of reflection about the processes of teaching and learning, which must of itself be healthy whether or not computers play an increasing role.

## Dialogue

Expert systems seek to offer a dialogue with the user in a way that has not been the case with previous systems. Here we wish to address the question of dialogue in the classroom, catalysed by the consideration of expert systems.

Marek Sergot and Peter Hammond of Imperial College have developed the Query The User Facility as an extension to logic programming, and provided the enabling software in their APES system. The design principle which they invoked was that of symmetry between the system and the user: as a heuristic device they propose that both the system and the user should be assigned the same logical status. Each can ask and answer questions, seek and give explanations. At the end of an interaction the sum total of knowledge of the user and the system may be the same, but new conclusions will have been reached as a result of sharing and manipulating the knowledge.

One interesting corollary of this approach is that in many respects it ceases to matter whether we are seeking advice from, or giving information to, a man or a machine. Many of the issues concerning expert systems dialogue, on this view, largely have to do with describing problems of knowledge in a manner that has been of interest since the Socratic dialogues were recorded by Plato. The machine drops out of consideration, leaving the real problems, as they always have been, concerning knowledge and its structures.

In considering a dialogue with an expert system, or with an expert, we have to give attention to the level of language used, and the assumptions that are made by each about the knowledge of the other. We are likely to want an approach based on mixed initiative, rather than either the 'teacher' or the 'student' asking all the questions. We use quotation marks here, because experience suggests that a very effective way of learning about a subject can be to teach it to another person or to a computer. This can be another way of characterising knowledge-based programming in education.

## Interaction

In different circumstances in eduction and training interaction is going to take different forms. If straightforward symmetry between the user and the system is built into the system, there may be a free flow of question, answer and information exchange driven by need. If the system is deemed to be the expert, it may be given precedence over the user, and there may be attempts to bring the belief system of the user into consistency with that of the expert system. Alternatively the user may be assigned a dominant role, in the driving seat and able to call on the services of a number of servant systems.

The same program and users may give rise to a great variety of interactions, depending on context and needs. We clearly have to take account of the institutional status and position of the people concerned in seeking to introduce innovatory approaches to computer technology. Unless the computer interaction is consistent with the desired human interaction it is unlikely to be successful.

## Explanation

Complete explanation by expert systems is a chimera. At present what is offered in the name of explanation tends to be a combination of redescription in terms of the rest of the knowledge base and a trace of the operational behaviour of the system. Very rarely does the system have a model of the user's belief system, with which it is seeking to be reconciled through interactions.

Explanation stops when we know how to go on — the explanation that is required says as much about the user and his or her model of the problem as about the system.

## Some expert systems

Given here are several expert systems devised by staff and students at Kingston College of Further Education. Their variety may prove surprising at first glance, but in each instance it should be recalled that a specific problem is being addressed. Often they have been conceived by staff as aids to instruction, like the following one which distinguishes between plant and animal cells, a perennial problem in Biology.

```
advice cell is eukaryotic if
   mitochondria present

advice cell is prokaryotic if not
   mitochondria present

advice cell is plant if
   cell is eukaryotic and cellulose cell
   wall present

advice cell is animal if
   cell is eukaryotic and not cellulose cell
   wall present

advice cell is bacterium if
   cell is prokaryotic and murein present
   and not phycocyanin pigment present
```

```
advice cell is blue-green alga if
   cell is prokaryotic and not murein
   present and phycocyanin pigment present
```

Another scientific example is a simple piece of advice on the mechanisms operating within diabetic insulin control regimes. It states:

```
advice
   blood glucose concentration decreases
   due to increased
   uptake/respiration/conversion to
   glycogen

if
   insulin increases

advice
   blood glucose concentration increases
   due to reduction in
   uptake/respiration/conversion of
   glycogen to glucose by cells

if
   insulin decreases

advice
   increased blood glucose concentration
   will decrease to set point

if
   blood glucose high and insulin increased

advice
   decreased blood glucose concentration
   will increase to set point

if
   blood glucose concentration low and
   not insulin increased
```

For the purposes of students taking courses in electrical engineering a fault diagnosis system can be used to introduce the concepts of electronic circuits.

```
reading is not steady as key is held still if
   the galvanometer is over sensitive
```

meter gives constant non-zero reading *if*
circuit is not ordered correctly

meter gives constant non-zero reading *if*
rheostat is too large

balance is at end of wire *if*
circuit is not ordered correctly

balance is at end of wire *if*
cell 1 / 2 is discharged

balance is at end of wire *if*
driver cell is too large

balance is at end of wire *if*
rheostat is too large

balance is at end of wire *if*
driver cell has lower PD than 1 / 2

cell 1 / 2 is discharged *if*
test meter reads > 1 V

circuit connections are loose *if*
terminals are not properly tightened

circuit connections are loose *if*
resistance across junctions is infinite

Even more obvious, though logically impressive, is a central heating fault diagnosis system. It runs:

problem is no heat *if*
problem is no heat in one area

problem is no heat *if*
problem is no heat throughout

problem is no heat in one area *if*
problem is no heat in radiator

problem is no heat in one area *if*
problem is poor insulation of that area

```
problem is no heat in radiator if
   problem is air lock in radiator

problem is no heat in radiator if
   problem is air lock in supply pipe to
that radiator

problem is no heat throughout if
   problem is all radiators not hot enough

problem is no heat throughout if
   problem is poor insulation throughout

problem is all radiators not hot enough if
   problem is air lock in main supply pipe
   to all radiators
```

A fault diagnosis system for use in a creamery was produced by an electrical engineering student as a project. Again it reveals the value of knowledge-based programming. The logical structure was a part of the learning process itself.

```
cream not at outlet if
   milk not at balance tank

cream not at outlet if
   total desludge activated

cream not at outlet if
   milk pump not running

cream not at outlet if
   cip bend still connected

cream not at outlet if
   hand screw in too tight

cream not at outlet if
   fault on compomaster

cream wrong fat % if
   cream selection wrong on compomaster

cream wrong fat % if
   high back pressure on skim side
```

```
cream wrong fat % if
   milk fat % incorrectly defined

cream wrong fat % if
   incorrect parameters on compomaster

milk not at balance tank if
   3-way cock in off position

contact electricians if
   electrical fault exists
```

Less technical, though no less rigorous, is a simple and humorous computer dating example used to introduce Youth Training Scheme and Certificate in Pre-vocational Education students to the concepts of an expert system.

```
he / she is a suitable match if
   you have similar tastes and
   you have similar interests and
   you have similar characters and
   you find him / her attractive

he / she is fairly suitable if
   you have similar tastes and
   you have similar interests and
   you have similar characters

he / she is fairly suitable if
   you have similar tastes and
   you have similar interests and
   you find him / her attractive

he / she is fairly suitable if
   you have similar tastes and
   you have similar characters and
   you find him / her attractive

he / she is fairly suitable if
   you have similar interests and
   you have similar characters and
   you find him / her attractive
```

A general application concerns safety, especially in science laboratories. The clarity of the exposition has been found helpful and reassuring. Here is a portion of the expert advice.

```
advice Use a water extinguisher if
   Type of fire is Class A and
   site is away from electrical
   installations and residue is not a
   problem

advice Use a sand bucket if
   Type of fire is Class D

advice Use a powder extinguisher if
   Type of fire is Class B and
   residue is not a problem and
   site is near to electrical
   installations and
   fire is in enclosed space

advice Use a powder extinguisher if
   Type of fire is Class A and
   site is near to electrical
   installations and
   residue not a problem and
   fire is in enclosed space

advice Use a vaporising liquid extinguisher
   if
   Type of fire is Class B and
   residue is a problem and
   site is near to electrical installations
   and
   fire is not in enclosed space

advice Use a vaporising liquid
   extinguisher if
   Type of fire is Class A and
   site is near to electrical installations
   and
   residue is a problem and
   fire is not in enclosed space
```

As a final example, this expert system assists business studies students in understanding legislation by explaining planning procedures.

```
planning permission is required if
   small building is to be constructed and
   works are not for benefit of house
   occupants
```

```
planning permission is required if
   small building is to be constructed and
   works will occupy more than half the
   garden

planning permission is required if
   small building is to be constructed and
   works will require new access to a public
   highway

job is general maintenance if
   job is redecorating outside

job is general maintenance if
   job is repointing

job is general maintenance if
   job is putting shutters on windows

job is general maintenance if
   job is building bay window and
   window will not extend beyond any wall
   towards road
```

## 3.5 Intelligent tutoring systems

One outcome of the fascination with computer technology in recent years has been the suggestion that a great deal of the work of teachers could be taken over by *intelligent tutoring systems*. Though the notion may seem very attractive, and indeed it is currently being pursued by large numbers of researchers and larger amounts of money, there remain tremendous problems in building even a modest intelligent tutoring system. Only a brief glance at good teaching, and all that goes into it, is enough to indicate the range of difficulties involved. Firstly, there is the body of knowledge itself; this has to be passed over to the student, with all its meanings, applications and analogies fully explained.

The second problem is ensuring that the student understands what he has learnt. Teaching and learning have never been synonymous. Without interaction between learner and teacher it is unlikely that any views the former may have on what is being taught will be sufficiently modified to overcome any prejudices blocking full comprehension. Teaching strategies are constantly changed when learning difficulties appear.

This is not to say intelligent tutoring systems are utterly impossible. Rather it is sensible to see them as aids to specific educational or training problems. Such an approach informs the Industry and Commerce Group at Kingston College of Further Education. Its staff draw on expert systems in order to tackle training difficulties in fields as diverse as tyre retreading, bet settlement and insurance procedures. At present, however, it appears that engineering offers the most fruitful area for intelligent knowledge-based systems, especially where machine management is concerned.

## Teaching strategies

Even if we can agree on the essential knowledge of a subject that needs to be acquired by a student, experienced teachers and lecturers will be aware that there is a wide choice of strategies available which can be deployed to aid in the process of transfer, acquisition and learning. For an intelligent tutoring system to make a useful contribution to education and training it will have to have knowledge of these strategies, and the capacity to select between them during the teaching process, then putting them into practice.

Intelligent tutoring systems suffer from a fundamentally disabling flaw. Not being humans, they cannot themselves perform the tasks they are teaching, they do not know what it is to learn except in a formal and impoverished sense. Our argument is that these observations should not lead us to wholly discard intelligent tutoring systems, but to locate them in their proper subordinate roles.

The consideration of teaching strategies in order to develop an intelligent tutoring system is itself a healthy and worthwhile activity. Too often teachers are assumed to know what they are doing without ever being obliged to reflect on the matter. We cannot point to ready-made gospels of good practice, infallible guides to subject teaching method and classroom management. We can, however, note that some teachers and lecturers, given a degree of planning and preparation and some thought about the process of teaching they are going to undertake, can generate extremely effective learning experiences for their students. To the extent that the students learn something in a lesson or course it can be said that the teacher has taught something — but it is rarely tightly specifiable or quantifiable.

In carefully defined subject areas, with clearly specified teaching strategies, and a high level of understanding of the level and needs of the individual student, intelligent tutoring systems can have an increasing role. Though at considerable expense of time, money, and above all, thought. The subject areas, by virtue of their danger, rarity, expense or cost-benefit characteristics, will need to be carefully chosen, together with clearly specifiable and quantifiable intended performance outcomes to be fixed as criteria against which the performance of the system can be judged. Advanced medical, military and scientific applications could be appropriate, given a sound research basis in advance of the use of the system.

### Understanding the student

The major conceptual barrier, which can be eroded but not conceivably removed, is that of understanding the student. The philosophers have explored the problem of 'other minds', and concluded that there are clear limits to the extent to which we can fully understand and represent the workings of other people's minds, and limits to the extent to which we can describe our own innermost thoughts and feelings. All of these findings carry over even more so to the world of intelligent tutoring systems, and suggest that perfect systems are out of the question, as indeed is the idea of a perfect teacher.

As experienced teachers and lecturers have learnt, largely by personal trial and error, teaching is not a one-way process. We have to start by establishing a relationship with our students, by finding a common language and vocabulary with which to work. Each situation is different, though much experience can be carried over and applies, with changes and adjustments, in new circumstances. The most flexible general-purpose problem-solving system is the human brain, which can be augmented by a choice of input and output devices to aid the process of communication - including the eyes, ears, nose and hands.

Intelligent tutoring systems have to date been constrained by their interface with the human user, which has tended to be in the form of glass teletype — an interaction between typewriters. Advances in graphics and object-oriented programming environments promise great improvements, but enormous deficiences will remain compared to the human communicator. An intelligent tutoring system will not be able to make jokes about the previous day's football match, or the romantic involvments of the different students. It cannot enjoy a glass of wine or a cup of coffee after a hard lesson, or meet in a pub to discuss the subject less formally. It is unlikely to be able to turn itself off if it becomes boring, and choose a more appropriate subject to deal with on a sunny summer Friday afternoon. How then can it hope to understand us or our students?

## 3.6 The impact of the 1990s

Futurology is an uncertain science. In the past the predictions made about the development of new technology have been invariably wrong, or at the most only partly correct. In the late 1940s IBM believed that the world demand for computers would be half a dozen large-scale systems. The problem is of course the inability of anyone now to see the full implications of future invention and application. For this reason we have chosen to point out here the general direction advanced information technology will be likely to take during the 1990s rather than dwell on specific areas of development.

## Technological change

Nonetheless, we are living in a period of rapid change in computer technology. Even without the Japanese Fifth Generation Computer Project it is becoming obvious that our daily lives are being influenced by the spread of enhanced microcomputer facilities. The vast amounts of money being spent on fifth generation computer research and development in Japan, the United States and the EEC should produce a quantum leap, however. New machines capable of working in parallel ought to emerge soon as viable computer systems. Already at Imperial College of Science and Technology the prototype of such a machine is in operation; it is known as ALICE. In the same manner the work currently being done on VLSI technology, flat screens, and new programming environments and languages will produce a wide range of computing benefits, though perhaps not those we might immediately anticipate. But it would not be an unreasonable assumption that together they will allow the manufacture of more powerful and more compact computers. These new parallel computer systems could well employ thousands of processors in problem-solving and allow computing to move entirely beyond the limitations of the von Neumann machine.

Even more dramatic could be the attempted revolution in communications between machines and human beings, which aims at a friendly and easier dialogue. Just as we have become accustomed to having our own telephone, so in the next decade we may take it for granted that a computer will have a telephone line of its own so that it can dial up automatically remote sources of information, remote data bases, and down-load that information into our personal computers. In this way machines may become the most powerful method of communicating with business and social colleagues. Instead of telephoning them, one is as likely to write a letter and send it automatically to their computer system. Possibly the word 'letter' will be an anachronism because such communications may never be written down on paper.

The amount of information that we are all going to have to handle, learn to sift, or to ignore, grows as rapidly as the pace of technological development. By linking computers with the telephone network we are already in a situation where businesses can access billions of pages of information from remote databases, including every word from the *New York Times*, The *Guardian*, and *Pravda*. While it is unlikely that this sort of information will be of interest to us in our own homes, other media will offer a comparable explosion of personal data and could easily overwhelm us. The video disk, originally designed to bring cinema into the home, now provides a storage device for thousands of pages of text, pictures or photographs. The family album of the future could be viewed during the commercial breaks on television. The compact disk (CD), developed for studio quality sound, is being harnessed to allow cheap encyclopaedic storage facilities for personal computers in the form of CD-ROM.

We can be fairly certain that access to all this technology, these communication devices and long-term storage, will require that the computers be programmed in something more powerful than BASIC. Will it be PROLOG, LOGO or SMALLTALK? In many ways it seems unlikely that we have already hit upon the solution to the sorts of languages we will require for parallel computing. Indeed extensive research is going to produce the next generation of logic-based or object-orientated languages that will allow us to make full use of the power of parallel machines.

## PROLOG in parallel

In the United Kingdom, Israel and Japan, research groups have designed two successors to PROLOG tailored to parallel computers: Concurrent PROLOG and PARLOG (PARallel LOGic). PROLOG was designed as an efficient implementation of logic on the sequential von Neumann machine. The shortcoming of this computer language remains its reliance on the strict behaviour of the underlying machine. If we change the machine, incorporating perhaps a thousand processors, then these new versions of PROLOG need to reflect the radical shift in operation that becomes possible. The solving of the problem has to be shared out between processors. Pure logic, although a powerful language for describing the problem, does not contain strategies for distributed problem-solving. PARLOG and Concurrent PROLOG combine the logic of a problem with additional information which enables the computer to process in parallel. The additional information comprises comments to the machine about the sort of strategies that should be employed to solve the particular problem.

These changes in machines and programming languages will affect how we all use computers, though only indirectly. Users of automatic cash dispensers and those who receive their monthly pay cheque from a computer are unconcerned about the language that has been used to write programs that do these jobs. What concerns them is the ease with which they are able to interact with such systems, and in this respect, the 1990s will present a variety of improvements.

## The computing environment

Computers such as the Apple Macintosh have shown that it is possible to make a computer system where the environment of use becomes friendly and the barriers between humans beings and computers are broken down. This should continue for all levels of computer use — the casual user perhaps in a public library seeking a piece of information or indeed the programmer, who is likely to be provided with more and more powerful working environments for producing better and better programs. Such environments will consist of tools for checking that a program is correct, for watching it work, and for identifying ways in which the computer could be speeded up to solve the problem more quickly. We are also likely to see expert systems that advise people about how to improve their programs.

Producing better programs is only possible when the problems involved are better understood. The work that is currently being conducted in the field of expert systems is very tentative. Researchers, programmers and developers are not sure how to write 'expertise' in the form of a computer program. Methodologies are required and expertise is needed, so that this process becomes more successful. With better tools, with more powerful machines, and with some of the limitations of the current technology removed, we may see the next generation of expert systems tackling more socially useful problems than those attempted so far.

## Changes in education

If the crystal ball of new technology looks a bit misty, predicting the changes that are likely to occur in the education and training world in the next 10 to 20 years may prove even more difficult. Buffeted as it is by political and social pressures, technology can only complicate the process of change. On a small scale the rapid decrease in cost of new technology is likely to mean that in the schools and colleges of the future the issues about whether computers will or will not be used in the curriculum are likely to disappear. It will be left up to the individual teacher or lecturer to decide whether to integrate the use of computers within the particular lesson or subject that they are teaching. Cheap but highly powerful 16-bit machines have now arrived, and while it is interesting to look ahead 10 years, the next two may provide to be the most decisive in terms of the uptake of powerful technology in the educational field.

This presents considerable opportunities for experimentation with styles of education. It is to be hoped that the days in which the Computer Studies Department feel that if the computers are not being used in every lesson they are misusing their resources, will soon be numbered. Staff and students are not compelled to watch every hour of educational television that is produced. Similarly they must not feel that a switched-off computer is a technological sin.

The widespread use of computers will doubtless have other

implications for teaching. Perhaps they will increase the amount of study done at home, dramatically extend the current age range of students, or alter the range of topics and subjects that can be taught.

Could all this signal the death of computer studies and computer literacy classes? It is to be hoped, particularly in secondary and further education, that a familiarity with technology will become second nature to most pupils and students. Although some remedial instruction may be needed for a small number of young people, we do not have to insist that every student receives something like computer writing exercises. Similarly, we can only hope that in the next few years the phasing out of computer science as a special subject for all but a few pupils will take place. Computers must be viewed as an integral part of everything that is learnt, lest computer science and computer scientists be placed on a pedestal. It may certainly seem controversial to suggest that the child who spends hundreds of hours programming a home computer in his or her bedroom is to be discouraged, but it is important to realise that the successful computer programmers of the future are those who will be able to explain succinctly to a computer system what their problem is, rather than those who start with a program and spend weeks or months tinkering with it.

Industrial commercial training may well be transformed in similar ways. It is certain that some of the goals of intelligent tutoring systems will be realised for small skills areas in the next few years, using the resources offered by interactive video and CD-ROM. We can also expect a large amount of commercial training to be conducted by means of new technology.

## Attitudes to change

Change is uncomfortable. Most of us would prefer to think that many of the things that we know and understand will remain constant. It is uncomfortable for the computer science teacher that the subject he learnt two or three years ago is now out of date. It is uncomfortable for the purchasing department that the computer system which they have just bought will be obsolete in three years time. It is uncomfortable that the database or word processing package that has recently been purchased has already been superceded. It is uncomfortable for some politicians that the process of education cannot be computerised. It is increasingly uncomfortable, indeed many would say indefensible, that not enough attention is being paid to the effect that new technology is having within society.

Would it not be better to go back to a time when technology had not been invented? Would it not be better to call a halt, at least for the time being, and plan the social and economic strategies that will put the new technology successfully in place without hardship? New technology is unavoidably about change, however: changes in hardware, in software, in style of use, but even more so about changes in attitude. It may not be comfortable, but what we must do in education and training is adapt people to change; we should now be preparing a generation not of specialists but of adaptable people who are able to modify their actions and jobs in the light of what is happening around them. New technology increasingly offers new opportunities for training. These opportunities must be taken when and wherever possible. The middle-aged pipe fitter who has worked all his life in the steel industry will be obliged to retrain for another field of work. Yet this cannot reasonably be expected to occur without the economic responsibility being shouldered by society for him, his family and his environment.

Fifth generation computer systems are an ideal, towards which the world is now progressing. As a piece of high technology they are exciting, but without proper attention being paid to their social, educational and economic implications they will become not only meaningless but disruptive. Education as ever is in a pivotal position in respect of this profound technological transformation. In the United Kingdom teachers have already begun to come to terms with the role they will need to play in preparing society for its impact, a move which must be to our national advantage when compared with the divide which persistently separates research from education in countries such as Japan. Although other nations may easily outstrip us in producing the machines and the software that we use, we have ourselves to face up to our responsibilities in determining the way in which there machines will affect our social life. It matters far less whether there is a fifth generation of computer systems than whether the present and next generations of students are in a position to benefit from their greater efficiency.

# 4 A Practical Approach to Getting Started

## 4.1 Describing knowledge with facts and rules in MITSI

Logic programming is no longer the preserve of the research laboratory: since 1980 implementations such as micro-PROLOG (from Logic Programming Associates) have been available for personal computers and have been in use in education and training with non-specialists in the United Kingdom and internationally, with many projects linked by the PROLOG Education Group. At Kingston College of Further Education, for example, an Information Technology Development Unit has been established, based on the use of logic programming and knowledge-based tools, which works on collaborative research and development with industry and education. The syntax used to introduce logic programming into courses at the college is that of 'MITSI' (Man In The Street Interface), developed by Jonathan Briggs. It does not differ significantly from PROLOG syntaxes covered by the new emerging international standards, but has some added facilities, such as simple explanation, for tutorial purposes.

One attraction of logic programming is the way in which the use of logic to represent knowledge can be combined with the use of logic for inference and a friendly man-machine interface, to supply the principal components of an expert system. The following is an example of a student assignment.

*Family relationships*

The basic unit of society is the family: we have all had a father and a mother, and we can often trace our ancestry from official records. We talk of the 'nuclear family' as being made up of parents and children, but many other family relationships can be important. In many societies we can talk of the importance of the 'extended family', including in particular grandparents. Sociologists, social historians and social anthropologists seek to understand family relationships to gain insights into the life of a society.

The purpose of this assignment is to explore family relationships using PROLOG. You can either describe your own family or another family that you know (perhaps even the Royal Family). We want to be able to extract the maximum information from the minimum number of basic facts, by defining new relationships in terms of existing ones.

Let us start by using the relations 'parent-of' and 'sex', which can provide us with all the facts on which we can base our rules. Here are some sentences about my family:

```
››› Arthur parent-of David.
››› David parent-of Richard.
››› Richard parent-of Robert.
››› Robert parent-of Chris.
››› Bobbie parent-of Chris.

››› Arthur sex male.
››› David sex male.
››› Richard sex male.
››› Robert sex male.
››› Chris sex male.
››› Bobbie sex female.
```

1. Type in corresponding sentences about your chosen family. We can now define some further relations, using variables. Variables are preceded with '_' and stand for an unknown individual. We can say, for example, that a person's mother is the person's female parent:

```
››› _A mother-of _B if _A parent-of _B and
      _A sex female.
```

2. Type in this rule and then add a rule defining the relation 'father-of', in a similar manner.
Test your rules by asking PROLOG questions of your program, for example

```
››› _A father-of _B?
```

and edit your facts and rules until you are content with them.

3. Using the facts and rules above, add rules defining the relations
(a) 'child-of'
(b) 'grandparent-of'
(c) 'grandfather-of'
(d) 'grandmother-of'.
Test and modify your rules by asking questions and editing as before.

4. You should now expand your program by adding information about some more family members, and then add a sentence describing the sex of each new family member.

5. Using the facts and rules in your program, add rules to define the following relations:
(a) 'sibling-of'
Your sibling has the same mother and father as you do, and is different from you. You can add a rule to define 'different-to' as follows:

››› _A different-to _B if not _A equals _B
(b) 'brother-of' (a brother is a male sibling)
(c) 'sister-of' (a sister is a female sibling)
(d) 'uncle-of'
(e) 'aunt-of'
(f) 'nephew-of'
(g) 'niece-of'
(h) 'cousin-of'.
Test and modify your rules as before.

6. You are descended from ancestors who lived many hundreds of years ago. Your ancestors are your parents, your parents' parents, and so on. Write two rules which will enable you to identify all the known ancestors of a person named in your program. Choose the easy case first:
››› _A ancestor-of _B if .....

7. Other family relationships may be particularly important to you. Add new rules to define them. (Examples could be: maternal-grandmother, male cousin, grandson.)

8. Ask questions of your system, adding further facts about members of your chosen family as necessary. You should find that you clarify both your knowledge of your own family and the gaps in your knowledge.

## 4.2 Tackling advanced information technology in further education

Expert systems are being put to diverse uses in industry and commerce: planning, advice, diagnosis, help, selection, configuration and training. In each case they manipulate knowledge, imitate human experts, answer questions, provide advice and explain their reasoning. In education they can be used as tools to think with, as a resource, as tools to teach with, as tutors and in administration. Too often, however, tools such as PROLOG, APES, ESP ADVISOR and Xi have proved hard to use, with complex syntax and confusing models of knowledge and of the user. They assume a degree of specialist knowledge and familiarity with computers which is not generally justified in further education institutions. What is needed is simple systems with which students and staff could start to explore new powerful ideas. Among the results of the project at Kingston College of Further Education are a 'starter pack' of such simple systems, with working examples and accounts of teaching experience.

Knowledge engineering can play a central part in learning, particularly when the student is seen as the knowledge engineer (in an extension of Piaget's approach to learning which has gained popularity in primary education). Expert systems ideas and tools can be used in modelling knowledge, researching the subject, refining knowledge and supporting discussion.

The use of expert systems by FE lecturers can have a number of implications, including new resources, new courses, enhancement of existing courses and new approaches to staff development. Expert systems can be very useful in training further education staff, as lecturers frequently have industrial experience, and teach in a work-oriented educational environment. FE colleges offer a bridge between education and industry, paralleling the kind of links sought in the Alvey Programme.

## 4.3 How to apply expert systems within your subject

There are two critical stages in gaining familiarity with expert systems. The first involves technical familiarity with the technology, and the second, arguably considerably more challenging, concerns understanding and beginning to realise the potential of expert systems applications within the curriculum.

### Gaining familiarity with the technology

Expert systems technology in the abstract has no practical value: it must be applied to real problems of knowledge. A good way of starting is to explore systems written by other people, just as before writing a book on a particular subject it is wise to review the available literature. Knowledge-based programs should make some sense when you read them, if they have been written in a clear declarative style. They also offer an advantage over books in that you can run programs to see what they are supposed to mean, and they should offer explanations of their working. This is the principle behind the 'Starter Pack' produced by

Jonathan Briggs, which offers several hours of exploration of the ideas, tools and applications of others on the basis of which the user can plan his own work.

In the pack, ADEX-Advisor and EGSPERT offer the opportunity of building new systems. Our experience suggests that it is a good idea to start with an idea of the general knowledge area to be represented. At first, hard and fast rules are not required: the objective is the acquisition of confidence in using the ideas and tools. Suggested areas and tentative rules for this initial foray could include:

1. *Advice on where to take your holiday:*

```
go to the Costa Brava if you enjoy the sun
                  and you do not object to
                  crowds of tourists
go to the Isle of Wight if you enjoy the sea-
side and
                  you do not want to go abroad
```

2. *Choice of transport for particular journeys:*

```
travel by train if you have to commute to the
city and
              there is no free parking at
              your work
travel by aeroplane if you have business to
do and
              your destination is distant
              and
              time is important to you
```

3. *Choice of clothing for different weather conditions:*

```
carry an umbrella if rain is forecast
wear an overcoat if the weather is cold and
               you have to go outside
```

4. *Choice of newspapers or television programmes:*

```
choose The Times if you are a top person
choose The Financial Times if you work in
finance and
               you want to follow financial
               results
```

5. *What to do in an emergency:*

```
dial 999 if there is a fire in your home and
        you cannot put the fire out at once
```

```
administer first aid if you witness an
accident and
                 you have training in first aid
```

In each case a small system involving only one or two rules at first can be written and tested, asking questions, obtaining explanations and adding further rules, thereby gaining insights into the workings of the expert systems shell and the style of reasoning which it supports.

## Selecting suitable knowledge domains

In industry, it is generally considered worth encoding expertise if the human expert activity lasts longer than half an hour and less than a day. If giving advice takes much less than about half an hour, the investment is seen as too costly. The process of writing down the knowledge is likely to become too complex if giving advice takes longer than a day.

In education, our costs are largely in terms of time. It is clear, however, that some areas of knowledge are too vast to be considered. The complete GCSE Geography or Physics syllabus is beyond our reach. We may learn some useful lessons from industrial applications by looking at the classes of problem which are addressed, which have analogues in different curriculum areas.

*1. Choosing a product*
We may be choosing a hairdressing preparation, an appropriate site for industrial development, or a candidate in an election.

*2. Giving advice*
This might concern careers, conduct of a mini-enterprise, health education or road safety.

*3. Diagnosing a problem*
We might be studying central heating systems, the workings of the British economy, or the performance of the local football team.

*4. Explaining a process*
The process could be chemical, electoral, legal or psychological

*5. Analysing data*
The data might be derived from local census figures, trade figures, or the text of a Shakespeare play.

It is of course important that systems to be interrogated by students do not seem trivial. If our aim in a particular lesson is to promote discussion and modification of the rules, even seemingly simple systems can be effective with appropriate supporting materials. One classic example is a set of rules defining social class, which can provoke discussion of the underlying concepts. These rules can then be applied to a database made up of census data or entries from trade directories.

Another concerns simple taxonomic systems for birds, animals or plants, where we are trying to draw attention to similarities and differences in our sample population.

Once we have grasped the potential of expert systems technology we can take particular curriculum issues and explore them with the computer. Richard Ennals has developed classroom examples for middle and secondary schools in his *Beginning micro-PROLOG* (1983) and *Artificial Intelligence: Applications to Logical Reasoning and Historical Research* (1985). Jonathan Briggs, working with the PROLOG Education Group at Exeter University School of Education, has developed software tools for teachers and pupils which have protected them from the necessity to learn any programming skills in a language such as PROLOG. This has supported imaginative classroom work in solving murder mysteries, researching the historic place names of Devon, writing simulations of voyages of discovery, and reconstructing details of classical Roman society. There is potential for much greater classroom use of powerful tools for thinking with the computer, taking advantage of more powerful 16-bit microcomputers and a suite of software developed with students and staff across the further education curriculum.

## Knowledge acquisition

So, we have expert system shells available which we can begin to use with small sets of rules. We also note a growing body of experience in using such technology in teaching and learning. How do we proceed within our chosen subject?

Knowledge must be acquired from somewhere, and then 'engineered' into the form of rules which can be typed into a system. Sometimes the knowledge will already be in a form that is easily accessible. It may already exist as a set of rules or regulations, a diagram or a decision tree. Often, however, it will exist in a much more diffuse form, in someone's head: can it be extracted?

Commercial developers of expert systems use a variety of techniques to extract knowledge from experts. One method is to conduct repeated interviews in which the 'knowledge engineer' asks the expert first to talk about how he solves a problem and then progressively refines this account. These interviews will attempt to find out what 'heuristics' or 'rules of thumb' are used by the expert. Alternatively, knowledge engineers can watch the experts at work, only asking questions when their decision processes are unclear. They can ask the experts to write down their knowledge in full detail. Another approach is to train the expert himself to use the expert system tools, cutting out the knowledge engineer. It is likely that a mixture of all these methods will be used in developing a complex system.

## Knowledge engineering

The field is new, and as yet there are few formal theories. We can offer some guidelines based on our experience.

1. You should have a clear idea of the educational role of your final system, so that knowledge is not simply added piecemeal.
2. Your expert system should be prototyped, testing early versions before too much time and effort is put into one approach.
3. Delay the final choice of a knowledge representation language (and thus expert system shell) until you have explored several alternatives.
4. You may find it better to develop a system collaboratively, gaining from the interaction and discussion.

# 4.4 Using the ADEX Advisor

ADEX-Advisor is a simple expert system shell designed for use in education, allowing advice systems to be developed quickly by students, lecturers and teachers. Selected knowledge or expertise is entered into the shell in the form of rules. ADEX-Advisor can then use these rules to provide advice. It will ask the user questions to complete its knowledge, and once advice has been presented, the system can be asked to provide explanation. It was decided to aim for simplicity in designing ADEX-Advisor, enabling educational users to make a practical start, but not offering all the features of a large commercial system.

To use the system, on a Research Machines Nimbus or IBM-PC compatible computer, the user simply types 'ADEX' and presses return. The screen then appears with a menu of commands, which enable the user to load an existing program by using two keystrokes, one to select the command, and the second to select the program.

The first example program concerns planning permission. When you own a house and wish to build on extra rooms, build in the roof or add other features, you need to consider whether planning permission is required. These regulations are contained in a booklet available from your local council planning department. Similar representations can be made of regulations on supplementary benefit, family income supplement or industrial training grants.

By typing 'd' we can display the rules, such as:

```
RULE 1
advice
contact the local planning department to
   obtain planning permission if planning
   permission is required
```

```
RULE 3
advice
planning permission is required if
    small building is to be constructed and
    works are not for the benefit of house
    occupants
```

If we quit from this menu by typing 'q', and then ask for some advice by typing 'a' and 'a' again, the computer will try to oblige. ADEX-Advisor looks through the rules to see if any of the advice is applicable, and asks the user questions to help it decide. The first question is again in the form of a menu, from which we choose the answer or answers which are applicable:

```
You intend
    to build a garden shed
    to build a sun lounge
    to build.....
```

If we are building just a garden shed, we can press the space bar to select that option, and then press return to indicate that this is the sole choice. ADEX-Advisor will then ask further questions, such as:

```
It is true that
    works are for the benefit of house
    occupants y/n?
```

to which a 'y' or 'n' answer is required. After a series of questions, advice will be offered, such as:

```
Contact the local planning department to
    obtain planning permission for this job
```

together with a further menu of options. We could choose to ask why this advice had been given, by type 'w', and the computer will begin to explain its conclusion:

```
because
    planning permission is required
```

Further explanation is given when any key is pressed. The user can then quit the menu by typing 'q' and exit from the system by typing 'e'.

Having used commands on the menus to clear the previous rules, the user can use the insert command (typing 'i') to enter a set of rules concerning his chosen subject. Richard Ennals has built a system to provide advice to lecturers on educational technology, teaching method and reprographics. It is made up of over a hundred rules describing the subject domain. For example:

```
teacher should consider teaching package if
   mode of delivery is distance
teacher should consider references if
   mode of delivery is seminar
teacher should consider handouts if
   mode of delivery is seminar

mode of delivery is distance if
   students are separated from teacher and
   there is a medium of communication and
   students receive support
mode of delivery is seminar if
   class size is 20 or less and
   subject matter is formal and
   students are encouraged to contribute
```

The consultation system is initiated by the user asking for advice from the system, using the menu on the screen. The system then asks the user questions, trying to establish which conclusions of rules hold true, and presenting them as advice. After each piece of advice more can be asked for. At a given stage, when a number of pieces of advice have been offered, the user may choose to recap advice received to date, together with explanations. For example, the user could be advised: 'discuss unit with team', 'check current resources in resource base', 'consider the objectives of the unit', 'assess the current skills of your students', 'consider handouts', 'use Macintosh OHP maker'. As new equipment becomes available its use can be described in additions and changes to the program using menu-driven editing facilities.

John Seaman has used the ADEX-Advisor system with classes of YTS and CPVE students, with an emphasis on catching the student's attention and imagination in the use of information technology. He has developed systems which advise on finding a partner of the opposite sex and on unfair dismissal from work, and encouraged classes to develop their own systems. Here is an extract from a student system:

```
advice
mend puncture if your bike has a puncture
advice
tighten cones if wheels are loose in
   holdings
advice
adjust brake nut to safe tightness if brakes
   are loose
```

He reported that 'the majority became quite interested in using such systems and proved quite adept at setting up simple versions or amending existing ones. They seemed quite happy to accept such systems as another source of information along with the library and their lecturers.'

Importantly the use of expert systems in the classroom provoked discussion of the subjects under consideration: 'On one occasion the system giving advice on unfair dismissal provoked quite a heated debate on the justice or otherwise of those rules.'

## 4.5 Using the EGSPERT Browser

When we first pick up a new book on a subject of interest to us, we do not necessarily read it from cover to cover. We are more likely to browse, skimming through to find points of interest, looking at particular sections in some detail with questions in mind, finding out whether the book puts forward a view we can share or find useful.

It is with this educational context in mind that the EGSPERT Browser was developed, allowing lecturers and teachers to explore the possibility of browsing expert system rules. As with ADEX-Advisor, knowledge or expertise is entered into the shell in the form of rules. EGSPERT then allows these rules to be explored, and 'why' and 'what if' questions can be posed. Again, the system has been designed for use on 16-bit microcomputers in schools and colleges, and is not intended to provide all the features of expensive commercial systems.

In order to use the system, the user types 'EGSPERT' and presses return. The screen then appears with a menu of commands, which enable the user to load an existing program by using two keystrokes, one to select the command, and the second to select the program.

The first example program concerns the faults which can occur when cooking a souffle. Is is not complete, and indeed can be extended by the user during the session. By typing 'd' we can display the rules, such as:

```
RULE 1
souffle is imperfect if souffle is black on
   top
RULE 2
souffle is imperfect if souffle is watery
```

In order to browse the system, we first type 'q' to quit the current menu, then type 'b'. At the top of the screen the phase

```
souffle is imperfect
```

will be displayed together with another menu. We can type 'w' to find out why a souffle might be imperfect, and EGSPERT will display a list of possible reasons:

```
souffle is imperfect because

   souffle is watery or
   souffle is too soft or
   souffle it too firm or
   souffle is black on top or
   souffle does not rise or
   souffle is collapsed
```

Using the up and down cursor keys we can select 'souffle is black on top' and press return. EGSPERT will provide reasons why the souffle might be black on top:

```
souffle is black on top because

   temperature is too hot or
   cooking time is too long
```

Similarly we can pursue why the cooking time might be too long, and receive the suggestion that the timer is faulty. We could then type 'i' for 'if' and EGSPERT will ask:

```
what may happen if timer is faulty

if timer is faulty then

   cooking time is too long or
   cooking time is too short
```

If we ask 'what if' again, typing 'i' and selecting 'cooking time is too short', we learn:

```
if cooking time is too short then

   souffle is watery or
   souffle is too soft
```

We can then ask 'why' a souffle might be too soft:

```
souffle is too soft because

   temperature is uneven or
   cooking time is too short or
   eggs are overbeaten or
   mixture is too thin
```

A final way of moving round the rules is to find similar concepts in the knowledge base. It is evident from the example so far that temperature is important to a successful souffle. Type 's' for 'similar' and select 'temperature is uneven', and EGSPERT will display all the parts of the rules which start: 'temperature is...'

```
temperature is uneven is similar to

    temperature is too hot
    temperature is too cool
```

With these commands and ideas the user can now either extend the current example or develop a knowledge base of his own.

John Stone was concerned to investigate problems in his subject, Physics, which required logical reasoning rather than just a breadth of factual knowledge. He chose to use the EGSPERT Browser to explore the use of a potentiometer. He found the exercise of developing the system an educational one in itself, testing his understanding of the scientific problem. He then used the system with a revision physics A-level class. He concluded

> 'Use of the system encouraged the students to take a more logical approach to solving problems they encountered.
>
> In a practical situation the emphasis often tends to be to make the circuit work rather than to investigate why it does. The series of suggestions offered by the system offered a wider selection of possibilities than a teacher would have time to select in a laboratory situation and yet offered less than the complete solution the student might demand from the teacher.
>
> The use of expert systems would seem to lend itself to the experiential learning approach and represents a considerable advance on the 'knowledge first, questions after' implicit in much of the currently available computer software.'

## 4.6 Using a knowledge base as a resource

We have earlier illustrated the use of the MITSI system to develop small knowledge bases. In many subejct areas larger knowledge bases can provide a powerful and flexible educational resource. Judith Christian Carter has developed a knowledge base for classroom use, concerned with vitamins and diet. Here are some of the rules from the knowledge base:

```
_food-Y contains _vitamin-X if
    _food-Y made-from _food-Z and
    _food-Z contains _vitamin-X
_food-Z prevents _some-disease if
```

```
    _food-Z contains _vitamin-X and
    _vitamin-X prevents _some-disease
_food-Z eaten-for _some-reason if
    _food-Z contains _vitamin-X and
    _vitamin-X needed-for _some-reason
```

Using MITSI we can then ask questions such as:

```
herrings eaten-for _reason?
potatoes prevents beri-beri?
_what contains vitamin-A?
```

and the answers can be explained.

There is a problem, however, for the uninitiated user of such a large MITSI knowledge base. In order to ask a useful question about the subject one needs to know the syntax of MITSI and how the knowledge has been structured in the knowledge base, the names and interconnections of the relationships. It was for this reason that a new system, Q, was developed to provide easier access to the knowledge base, making it available as a classroom resource. The same facts and rules which were entered in MITSI can be interrogated using Q, which is driven by a system of menus of commands, minimising the amount of typing needed (and the scope for error).

The user of Q-VITAMINS (which forms part of the Starter Pack developed by Jonathan Briggs), loads the system by typing 'Q-VIT' and pressing return. The screen then appears with a menu of commands. To ask a question the user types 'a'. Q-VIT will display a list of questions:

```
What foods should be eaten for _reason?
Why eat _food?
What foods help prevent _disease?
What diseases does _food prevent?
What vitamins are contained in _food?
What vitamins prevent_disease?
What diseases are prevented by _vitamin?
What vitamins are needed _reason?
Why does the body need _vitamin?
etc.
```

As before we can select by using the up and down cursor keys, and press return. If we select 'What diseases does _food prevent?' we are then presented with another menu:

```
all values for _food
 enter a value for _food
 choose a value for _food
 restrict value of _food
```

If we press 'c' the cursor will move to 'choose a value for _food' (following the initial letter). We are presented with a choice:

```
apple-pie
apples
bacon
baked-beans
bananas
beef
beefburgers
biscuits
bread
breakfast-cereals
and so on.
```

If we press return to select 'apple-pie', Q-VITAMINS will now answer the question:

```
What diseases does apple-pie help prevent?
```

with the following:

```
8 answers to this question

beri-beri                  keratomalacia
night-blindness            ostemalacia
pellagra                   rickets
scurvy                     xeropthalmia
```

Q-VITAMINS can explain its reasoning in terms of the facts and rules in its knowledge base. If we type 'w' for 'why', the system will display a list of the answers from which we can select one for explanation. If we select 'scurvy', we will receive the explanation:

```
apple-pie helps prevent scurvy because
   apple-pie contains vitamin-C and
   vitamin-C prevents scurvy
apple-pie contains vitamin-C because
   apple-pie is made from fruit and
   fruit contains vitamin-C
apple-pie is made from fruit is stated
fruit contains vitamin-C is stated
vitamin-C prevents scurvy is stated
```

## 4.7 Taking innovation beyond experimentation

It takes many years for educational innovations to take hold, and all too often practical experience in the classroom fails to live up to expectations. One problem is the common belief that the power lies in the technology, rather than in its appropriate use. Another is the assumption that a new technology or set of ideas should be fully consistent with, and judged by the same criteria as, the old. In educational computing we are currently undergoing a period of flux, of shifting paradigms, made less comfortable by the effective withdrawal of public sector financial support from pilot projects.

In 1980 micro-PROLOG was the only implementation of PROLOG available for 8-bit school microcomputers, and the syntax of the language appeared somewhat 'user-hostile' to the uninitiated. It was originally developed to support the teaching of logic as a computer language, with the emphasis on clear logical thinking in the classroom. In the following year PROLOG was identified by the Japanese as the starting-point for their new fifth generation of computer systems, and achieved sudden unaccustomed prominence. There are now numerous competing low-cost implementations of PROLOG for personal computers, and a diversity of applications to which they are put, including databases, expert systems, spreadsheets, natural language understanding systems, control and robotics. The British Alvey Programme of Research and Development in Advanced Information Technology, established in 1983 as Britain's response to the Japanese initiative, has placed considerable emphasis on logic programming and PROLOG, pursuing the objective of developing a new generation of software technology. Logic programming represented a change for computer scientists from their conventional procedural approaches and languages, and through a network of collaborative projects involving academia and industry considerable expertise has been developed, and a number of new logic-based British software products such as expert systems and expert system shells have reached the market.

The objectivies of education, however, are not the same as those of industry and advanced research. Educational success does not depend on the production of complete, perfect automatic systems to run complex industrial processes, nor do the software systems used have to be complex. The authors, with support from the Nuffield Foundation, have been involved in action research in the classroom, trying out and documenting experimental approaches with enthusiasm and commitment, leaving the process of evaluation to others who could aspire to more objectivity. We have not been teaching PROLOG, rather we have been seeking to use it as a tool to support the intelligent exploration of knowledge by students and teachers. In some senses the computer as such drops out of consideration: the focus is on educational exploration, in subject areas as diverse as history, classics, home economics, mathematics, physics and engineering. In the tradition developed in primary schools from the work of Piaget, we encourage participation and questioning.

Educational use of PROLOG in the United Kingdom has proliferated through the activity of enthusiasts, as is evidenced by the growing collection of introductory texts, and the experimental use of expert system shells such as Xi, Xi-plus, APES, and ESP-Advisor, all written in PROLOG.

There is now a sufficient literature and range of commercially available software on educational microcomputers to enable individual teachers and schools to make educational use of PROLOG. However, schools are frustrated by the lack of sufficient 16-bit microcomputers, and of specially commissioned educational software. More serious is the absence of funding for research and teaching, to build on the advances of the Alvey Programme for education and training.

## Innovation and institutional change

The hardest part of innovation is not the research and development of the technology, but the changes and adjustments that are required for individual professionals and the institutions within which they work. Among the issues that will have to be considered are: moves to team teaching; the introduction of desktop publishing; the effects of a core introductory course in advanced information technology offered to all students and staff; the role of students as consultants; the role of advanced information technology in the areas of multi-cultural education and special educational needs. In each of these areas it is not the technology itself which is the prime focus of interests, but the role that it may have in facilitating innovative activity in education and training.

# 5 Reference

## 5.1 Glossary

ADA A computer language developed at the behest of the US Department of Defense and named after Ada Lovelace, mistress of Charles Babbage, who is regarded as the first programmer. It combines many of the features from a large range of other computing languages and has been criticised for trying to be too many things at once.

ALGOL or algorithmic language, a high level computer language jointly developed by American and German scientists at the close of the 1950s. Despite being one of the first languages to be designed with a coherent structure, it has only had a limited commercial success.

ALICE A prototype parallel computer under development at Imperial College of Science and Technology.

Alvey Programme The British response to the Japanese Fifth Generation Computer Project, which was announced in 1982. Under the chairmanship of Mr John Alvey, Senior Director of Technology with British Telecom, the Committee was charged by the Department of Trade and Industry to look into advanced information technology and recommended a collaborative research programme between government, academics and companies. It identified four enabling technologies:

man-machine interface
intelligent knowledge-based systems
software engineering
very large scale integration (VLSI)

A further initiative in 1984 has addressed declarative systems architectures, supporting work on parallel computing, logic programming and large databases.

AM an expert system which handles mathematical concepts and explores the possibility of uncovering new rules or facts.

APES A suite of PROLOG programs that provide the PROLOG programmer with many of the features of expert systems.

Artificial intelligence (AI) The current effort to make computers capable of performing tasks which require intelligence when done by human beings. AI combines the full range of research topics including speech recognition, vision or image processing, games playing and robotics. The term is sometimes restricted to concentrate on new style of computer programming.

The focus of AI is symbolic or conceptual ideas rather than conventional numberical computation. The intelligent behaviour that AI programs now exhibit, which makes them different from numeric, database and word-processing programs, comes from the numeric, database and word-processing programs, comes from the ability to deal with symbols as concepts and ideas, not as a collection of meaningless symbols that only human beings cañinterpret.

ASK is an expert system devised by Hewlett-Packard in California. Its usefulness as an interface to a database is the ability to work on a microcomputer.

assembly languages or assemblers Notations for the representation of machine-code programs. They employ letters, numbers, symbols and a few words. Machine code is the set of binary numbers used in the programming of the operation of a given machine.

BASIC or beginner's all-purpose symbolic instruction code. An interactive computer language developed at Dartmouth College, USA in the mid-1960s to introduce novices to computing. In the Report of the Alvey Committee (1982) anxiety was expressed about the current preoccupation with BASIC in schools and it was recommended that the computer languages taught should be 'chosen with an eye to the future'. Perhaps if it had been called COMPLEX, BASIC would never have achieved its undeserved status.

COBOL or Common Business-Oriented Language, is used predominantly for commercial applications such as payrolls. It was intended to fill one of the gaps left by FORTRAN, another high-level computer language suited to the problems of Science and Engineering. Pressure from the US government helped to make it a leading language during the 1960s, when official policy declared that if a computer manufacturer did not offer COBOL with a machine, it would not be eligible as a supplier to the government unless it could demonstrate that it was offering a language that gave a better performance.

COMAL or common algorithmic-language. Initially devised for use in schools in Denmark, as an extension of BASIC, COMAL is now used in Britain as well.

compiler A piece of software which translates a program into a machine code which the computer understands. It compiles languages usually run much faster than interpreted languages.

database A collection of information stored in a computer. More recently the term 'knowledge base' has been introduced, particularly where knowledge is shared, as in PROLOG, in facts and rules. Databases are interrogated by means of a query language. Most databases, however, store information according to fixed formats and data can only be accessed in a restricted way. Relational database provide a more flexible approach.

declarative systems became part of the Alvey Programme for Advanced Information Technology in 1984. A declarative system is one in which knowledge is stored in a descriptive rather than a prescriptive form. That is to say, the 'what' rather than the 'how'. In the Alvey Programme research is underway to build a computer that is specifically designed to store and manipulate such declarative programs. Such a machine would be described as having a 'declarative systems architecture.'

ELIZA A programme designed to simulate the conversation between a patient and a psychotherapist. It reveals how easy it is for a computer to give 'an impression of intelligence'.

ESPRIT A programme in the EEC which has a similar brief to the Alvey Programme in Britain. The EUREKA initiative in France is also open to non-EEC members, however.

expert systems First developed in the United States, they are computer programs which were initially intended to replace human experts such as doctors, lawyers and tax advisers. They typically represent knowledge symbolically, examine and explain their reasoning processes, and address problem areas that require years of special training and education for human beings to master. In recent years the areas of expertise for which expert systems seem to provide solutions have become less high-powered. Their real benefits will be in assisting humans to be more productive.

fifth generation computers The name of the new computer systems presently under development in Japan and elsewhere in the world. The Japanese Ministry of Trade and Industry in 1981 decided to commission an extensive research and development programme aimed at producing computers which would be easier to use, less burdened by software generation, more reliable in performance and cheaper to run. The Alvey Programme is the British response to this technological challenge.

FORTRAN or formula translation, a high-level computer language developed in the United States between 1953 and 1958. Using a notation system reminiscent of algebra, FORTRAN represented an important advance over assembly languages, which made use of letters, numbers, symbols and such short words as 'add'. It is particularly favoured by scientists and engineers.

GOTO statement An instruction in a program that breaks its normal sequential flow. It is surprising that such a small word as GOTO should cause so many computer scientists so many problems. The current emphasis in computing is to produce programs that are structured, easy to use and easy to maintain. The GOTO statement in many early languages makes these goals very difficult to attain.

GUS is an expert system used in the booking of airline tickets.

integrated circuits are the implementation of a particular electronic circuit function in which all the devices required to realise the function are made on a single chip. *Very large-scale integration* (VLSI), one of the research and development fields in the Alvey Programme, represents an enhancement of the principle of integration. Specifically, this part of the programme is intended to ensure that by the late 1980s the United Kingdom has the capability to specify, design and test silicon chips containing a dramatically-increased number of interactive functions.

INTELLECT An expert system concerned with the interrogation of databases. Its designer, Intellicorp of California, markets expert systems and system-building tools for scientific and commercial applications.

intelligent front end A program which gives the user the means of making use of a complex software system, such as a statistical model or an extensive relational database.

intelligent knowledge-based systems (IKBS) This term is used in the Report of the Alvey Committee (1982) to cover work in databases, expert systems, inference and natural language understanding. In essence, an IKBS is a system which applies inference to knowledge to perform a task.

interpreters are pieces of software which translate sentences of a computer language into the specific machine code which the computer understands. Unlike a compiler, an interpreter will usually translate a portion of the program at a time — usually when the portion is being executed. Languages which are used interactively are very often interpreted.

LISP or list processing. Since 1961, LISP has been the most widely used computer language in artificial intelligence research. While it is not possible to run it on small computers, though not microcomputers because of the large amount of memory required, the decision of the Japanese to adopt logic programming probably spells its eventual decline and disappearance. However, in the United States where a large-scale approach to AI is still in favour, LISP remains very important. One of LISP's problems is that it exists in a huge number of different forms. These have names like FRANS LISP and GOLDEN LISP. Recent work has been directed at producing a standard form, known as COMMON LISP.

LOGO A dialect of LISP developed for educational use following the ideas of Seymour Papert. Along with PROLOG, it is the chief AI language implemented on microcomputers in schools in the United States and Britain. It is an excellent tool for introducing the concepts of procedural thinking and programming.

man-machine interface (MMI) This involves the interaction of man and computer, an important area of research in the Japanese Fifth Generation Computer Project. Considerable work is going on in Britain as a result of the Alvey initiative, including the development of voice-driven computer facilities.

micro-PROLOG is an implementation of PROLOG for microcomputers. Since 1980 is has been used for the project Logic as a Computer Language for Children based at Imperial College, as well as for research on expert systems, databases, and man-machine interface. Micro-PROLOG hads now been superceded by LPA PROLOG Professional. This is one of a large number of versions of PROLOG now available in the United Kingdom. Others include PROLOG 2, SD-PROLOG, and QUINTUS PROLOG, the last echoing the fifth generation in its name.

MITSI or man in the street interface. It was devised by Jonathan Briggs to explain ideas in logic programming and knowledge-based computing to non-specialists.

mouse A hand-held device by which the computer can be controlled, instead of using the QWERTY keyboard. Mice operate with Mackintosh and other advanced machines. One facility a mouse can control is 'windows' i.e it allows the user to divide the screen into sections for different purposes.

MYCIN An early expert system used to diagnose bacterial infections in hospital patients and to prescribe the appropriate antibiotic. Medicine appears to be a natural field for such computer applications and in the United States further interesting development work is in progress.

parallel computers Computers with several processors, or indeed thousands, working at once to solve a problem. ALICE, currently under construction at Imperial College, is a leading example. It is designed to handle HOPE and PARLOG, languages primarily developed for parallel processing.

PARLOG or parallel logic. A newly developed parallel implementation of logic programming.

PARRY An expert system concerned with abnormal behaviour, an area of considerable interest to artificial intelligence researchers.

PASCAL Descended from ALGOL, this computer language has been in very common use since the mid-1960s. It was in fact designed as a tool to assist in the teaching of programming.

PL/1 is a high-level computer language. In the early 1950s it was

National Physical Laboratory in Teddington, IBM changed its name from new programming language (NPL) to programming language number 1 (PL/1).

PLAN A program written in PROLOG to provide an environment in which children can write adventure games.

PROLOG or programming in logic, was developed in France by Alain Colmerauer in the early 1970s. It is based upon an academic tradition which for centuries has used formal logic to describe and represent human reasoning. In operation, PROLOG employs clear descriptive statements which are interpreted by the computer as computer programs. In many cases that description, or specification, will act as a program that solves the problem concerned. Its place in the history of computing is assured by its selection in Japan as the starting point for the core programming language for the Fifth Generation Computer Project, although we can confidently expect developments to replace it.

PROSPECTOR As its name suggests, PROSPECTOR is an expert system designed to aid geological exploration. It is implemented on INTERLISP, a powerful but relatively low-level dialect of LISP. Obviously the system is a large one because of the extensive field of knowlege involved.

QUEST An educational software package devised to exploit the historical and contemporary information held in records, directories and census returns. It has been written by AUCBE at Hatfield.

SHRDLU An early expert system with a remarkable range of language exchange. It has implications for robotics.

SMALL TALK A computer language developed in the United States that is especially good for graphics-oriented programming. It has common features with SIMULA, one of the first object-oriented programming languages. As its name suggest, SIMULA was originally intended for simulation work.

software engineering This term covers the entire range of activities used to design and develop software. It intends to replace the haphazard construction of computer programs with a more rigorous and more formal approach. In the Alvey Programme one of the prime objectives is the establishment of the United Kingdom as a world leader in software engineering by the late 1980s to early 1990s.

spreadsheets Automated financial ledgers that store data and allow the user to make projections of financial outcomes on a 'What if...' basis. They are helpful in financial planning and the communication of financial concepts.

TEIRESIS An expert system which helps transfer knowledge from an expert to a knowledge base. The system acquires new rules about the problem domain through an interaction that allows users to state rules in a restricted English vocabulary.

transputer The computer on a chip, developed by INMOS, and the building block for a number of new computer architectures.

von Neumann machine The fundamental principle underlying pre sent day computers was devised by John von Neumann in the 1940s. It is computing by means of a central processing unit, which with the assistance of a memory, handles a sequence of instructions through their examination one after the other. Fifth generation computer systems represent an attempt to move beyond this sequential limitation and more closely approach the complex thinking processes of the human mind.

XCON or RI An expert system used by the Digital Equipment Corporation to determine the configuration of computer systems.

## 5.2 Suggested reading and bibliography

ALVEY COMMITTEE REPORT (1982) *A Programme for Advanced Information Technology*, HMSO.

AYER, A.J. (1964) *Language, Truth and Logic*. Gollancz

BENSON, I. and LLOYD, J. (1983) *Information Technology and Industrial Change*. Kogan Page.

BODEN, M. (1977) *Artificial Intelligence and Natural Man*. Harvester.

BRIGGS, J.H. (1984) *micro-PROLOG rules!*. Logic Programming Associates.

BRUNER, J.S. (1966) *Man, a Course of Study* in *Towards a Theory of Instruction*. Harvester University Press.

BURNS, A. (ed.) (1984) *New Information Technology*. Ellis Horwood.

CLARK, K.L. and McCABE, F.G. (eds) (1984) *micro-PROLOG: Programming in Logic*. Prentice-Hall.

COTTERELL, A.B. (ed.) (1988) *Advanced Information Technology in the New Industrial Society*. Oxford University Press.

DAHL, O.J., DIJKSTRA, E.W. and HOARE, C.A.R. (1972) *Structured Programming. Academic Press.*

ENNALS, J.R., GWYN, R. and ZDRAVCHEV, L. (eds) (1986) *Information Technology and Education: the Changing School*. Ellis Horwood.

ENNALS, J.R. (1983) *Beginning micro-PROLOG*. Ellis Horwood and Heinemann Computers in Education.

ENNALS, J.R. (1985) *Artificial Intelligence: Applications to Logical Reasoning and Historical Research*. Ellis Horwood.

ENNALS, J.R. and COTTERELL, A.B. (1985) *Fifth Generation Computers: their Implications for Further Education*. Further Education Unit.

ENNALS, J.R. (1986) *Star Wars: A Question of Initiative*. John Wiley.

ENNALS, J.R. (ed.) (1987) *Artificial Intelligence*. Pergamon Infotech State of the Art Report.

FEIGENBAUM, E. and McCORDUCK, P. (1983) *The Fifth Generation*. Addison-Wesley.

FOUCAULT, M. (1970) *The Order of Things*. Tavistock.

FUCHI K. and FURUKAWA K. (1986) *The role of Logic Programming in the Fifth Generation Computer Project* in *Proceedings of 3rd International Conference on Logic Programming*, Shapiro E. (ed.), London 1986. Springer-Verlag.

GAINES, B. and SHAW, M. (1984) *The Art of Computer Conversation*. Prentice-Hall.

GILL, K.S. (ed.) (1986) *Artificial Intelligence for Society*. John Wiley.

GOODYEAR, P. (1984) *LOGO: A Guide to Learning through Programming*. Ellis Horwood.

HAYES, J.E. and MICHIE, D. (eds) (1983) *Intelligent Systems: the Unprecendent Opportunity*. Ellis Horwood.

MOREAU, P. (1984). *The Computer Comes of Age: the People, the Hardware, and the Software*. MIT Press.

NICHOL, J., BRIGGS, J.H. and DEAN, J. (eds) (1988) *Fifth Generation Computing in Education: PROLOG, Children and Students*. Kogan Page.

O'SHEA, T. and SELF. J. (1983) *Teaching and Learning with Computers*. Harvester.

PAPERT, S. (1980) *Mindstorms*. Basic Books and Harvester.

PARSONS, T. (1937) *The Structure of Social Action*. McGraw-Hill.

PASK G. and CURRAN S. (1982). *Micro-Man*. Century.

PIAGET, J. (1950) *The Psychology of Intelligence*. Routledge and Kegan Paul.

RAUCH-HINDIN, W.B. (1986) *Artificial Intelligence in Business, Science and Industry, Volume I, Fundamentals*. Prentice-Hall.

ROSS, P. (1983) *LOGO Programming*. Addison-Wesley.

RUSHBY, N.J. (ed.) (1981) *Selected Readings in Computer-Based Learning*. Kogan Page.

SAPIR, E. *Selected Writings in Language, Culture and Personality* (ed.).

MANDELBAUM, D.G. (1949) University of California Press.

J.J. SERVAN-SCHREIBER, (1980) *The World Challenge*. Collings.

TORRANCE, S. (ed) (1984) *The Mind and the Machine: Philosophical Aspects of Artificial Intelligence*. Ellis Horwood.

WALLACE, M. (1984) *Communicating with Databases in Natural Language*. Ellis Horwood.

WATERMAN, D.A. (1986) *A Guide to Expert Systems*. Addison-Wesley.

WINSTON, P.H. and PRENDERGAST, K.A. (eds) (1984) *The AI Business : Commercial Uses of Artificial Intelligence*. MIT Press.

YAZDANI, M. (ed.) (1984) *New Horizons in Educational Computing*. Ellis Horwood.

# Index